Economics for the Modern Built Environment

Recent decades have seen major social and economic changes across the developed world and consequent changes in the construction and property industries. The discipline of construction economics needs to respond to this. For instance, the importance of sustainable development has become recognised, as has the need to increasingly master the medium- and long-term consequences of construction, not only in the production but also in the management of buildings across their whole life cycle. And the new focus on the service rendered by buildings, as distinct from the buildings themselves, has prompted a new approach to the construction and property industries. Any economic analysis of these sectors has to take account of all the participants involved in the life cycle of building structures – not only in the design and construction, but also in the operation, maintenance, refurbishment and demolition of property.

This innovative new book draws on the work of the Task Group of the CIB (Conseil International du Bâtiment – International Council for Research and Innovation) on Macroeconomics for Construction. It pulls together discussions of mesoeconomic and macroeconomic models and methodologies in construction economics and presents an exciting approach to the analysis of the operation and function of the construction and property sector within the economy. Graduate students and researchers will find it an invaluable work.

Les Ruddock is Professor of Construction and Property Economics, and Associate Dean for Research in the Faculty of Business, Law and the Built Environment at the University of Salford. He is the Co-ordinator of the CIB Task Group on Macroeconomics for Construction.

Economics for the Modern Built Environment

Edited by Les Ruddock

Taylor & Francis
Taylor & Francis Group
LONDON AND NEW YORK

First published 2009
by Taylor & Francis
2 Park Square, Milton Park, Abingdon, Oxon OX14 4RN

Simultaneously published in the USA and Canada
by Taylor & Francis
270 Madison Avenue, New York, NY 10016

*Taylor & Francis is an imprint of the Taylor & Francis Group,
an informa business*

© 2009 Taylor & Francis

Typeset in Sabon by Swales & Willis
Printed and bound in Great Britain by
CPI Antony Rowe, Chippenham, Wiltshire

This publication presents material of a broad scope and applicability.
Despite stringent efforts by all concerned in the publishing process,
some typographical or editorial errors may occur, and readers are
encouraged to bring these to our attention where they represent
errors of substance. The publisher and author disclaim any liability,
in whole or in part, arising from information contained in this
publication. The reader is urged to consult with an appropriate
licensed professional prior to taking any action or making any
interpretation that is within the realm of a licensed professional
practice.

British Library Cataloguing in Publication Data
A catalogue record for this book is available from the British Library

Library of Congress Cataloging-in-Publication Data
Economics for the modern built environment / edited by Les
Ruddock.
        p. cm.
    1. Construction industry.   2. Construction industry—
Management.   I. Ruddock, Leslie, 1950—
    HD9715.A2E285 2008
    338.4'769—dc22   2008018360

ISBN10: 0-415-45424-7 (hardback)
ISBN10: 0-415-45425-5 (paperback)
ISBN10: 0-203-93857-7 (ebook)

ISBN13: 978-0-415-45424-7 (hardback)
ISBN13: 978-0-415-45425-4 (paperback)
ISBN13: 978-0-203-93857-7 (ebook)

Contents

Figures

Tables

Preface

The International Council for Research and Innovation (CIB) Task Group *Macroeconomics for Construction* (TG56) and Working Commission *Building Economics* (W55) provide international focal points for the discussion of economic methodologies for the study of the built environment sector. The work of TG56, in particular, is concerned with the production of economic information on the state of, and the development of models for the analysis of, the sector. The idea for this book arose from a workshop of TG56 where a proposal was made that researchers from the areas of construction and property economics should put together a book in which new developments in the field could be presented.

Getting to the point of producing a completed text has been an interesting process and, as editor of the book, the challenge faced was in finding suitable writers for each of the chapters. The 'best' people are usually the busiest! I believe, though, that I have been fortunate in getting together such an impressive team of contributors, with the necessary expertise to provide such interesting papers. These contributors are drawn from a variety of countries and constitute a truly international set of experts.

Maybe, the 'modern built environment' is an expression, which needs to be explained. Construction activity has changed in response to new demands over recent decades and a new approach to the role of the construction and property sectors is needed, in order to consider the role of the sectors in meeting the dynamic needs of the economy and of society. This book incorporates economic analysis of the broad construction and property sectors and these can be referred to collectively as the 'built environment sector'.

The aim of the book is to provide a structured set of material, which explores economic issues pertinent to the built environment and focuses on integrating models with real world applications, often in the context of case studies. Each of the contributions is relatively free-standing but, due to the themed nature of the text, there are obvious linkages between the chapters.

The book will assist researchers and students of the built environment to better understand the economic background to their field of study and, for the general academic community, practitioners and anyone else with an interest in the 'modern built environment', it is hoped that the ideas and directions in this work will provide inspiration.

Finally, I wish to take this opportunity to thank all authors of papers for their contributions to the book. Without their high level of commitment and support, it would not have been possible to produce this work.

Les Ruddock

Contributors

Editor's details

Les **Ruddock** is Professor of Construction and Property Economics in the School of the Built Environment and Associate Dean for Research in the Faculty of Business, Law and the Built Environment at the University of Salford, where he was formerly the Director of the 6* Research Institute for the Built and Human Environment. He is International Co-ordinator of the CIB Task Group on Macroeconomics for Construction (TG56) and has written extensively on the economics of the construction and built environment sectors. He is a past editor of the RICS Research Paper Series and sits on the editorial boards of several major journals.

Details of contributors

Jan **Bröchner** graduated with a MSc (C.E.) from the Royal Institute of Technology (KTH) and, receiving a BA from Stockholm University, he completed his PhD at KTH with a thesis on economic aspects of housing rehabilitation. Having worked for a large Swedish construction contractor, he has also been a research manager for the Development Fund of the Swedish Construction Industry (SBUF). In 1998, he was appointed to the Chair of Organization of Construction, at Chalmers University of Technology in Göteborg. He has published a number of articles in international journals and book chapters, mostly with a basis in industrial economics and often dealing with facilities management.

Timothy **Michael Lewis** graduated from the University of Liverpool with a BEng and MEng by research in Civil Engineering. He then worked with Sir Alexander Gibb & Partners (consulting engineers). He obtained an MSc in Technological Economics from the University of Sterling, where he worked as a Research Fellow for three years before joining The University of the West Indies, where he was awarded his doctorate in 1990, and made Professor in 2005. He is currently Deputy Dean of the Faculty of Engineering for Distance Education & Outreach and Graduate Studies. He has published four books, three monographs and over 60 papers, mainly in the fields of construction economics and management.

Tullio Gregori is Associate Professor at the University of Trieste, Italy, where he teaches International Monetary Economics, Monetary Policy and Applied Economics. He graduated at the same university and got a Masters in Policy and Planning from Northeastern University, Boston, where he joined the Center for European Economic Studies. His research topic focused on impact analysis with econometric augmented input/output models. He has published several articles on input/output modelling and structural analysis. Currently his research fields include regional labour market analysis and open economy models.

Steven Ruddock is a Researcher in the Research Institute for the Built and Human Environment at the University of Salford. His research interest focuses on the economic appraisal and management of facilities.

Jorge Lopes has a PhD in Construction Economics and is Professor of Construction Economics and Management in the School of Technology and Management at the Polytechnic Institute of Bragança, where he is Head of the Department of Construction and Planning. He is on the editorial board of the *International Journal of Strategic Property Management* and is an active member of the CIB Task Group 56 (Macroeconomics for Construction). He has various research publications in the area of construction economics, stock valuation, housing markets and the legal framework of the construction industry. He has collaborated in the *Doing Business* project of the World Bank and is also an external consultant to the Road Institute of Cape Verde.

Geoff Briscoe is a Senior Lecturer in The Built Environment Department in the Faculty of Engineering and Computing at Coventry University, UK. His academic interests are in the areas of construction economics, management and finance. He is also an Associate Research Fellow in the Institute for Employment Research at University of Warwick, UK. Over the last three decades, he has published widely, including textbooks, research monographs and numerous journal articles. His main areas of published research are in construction labour markets and supply chain management. He has worked in a consultancy capacity for several UK construction organizations, including CITB-Construction Skills and the Institution of Civil Engineers.

Johan Snyman obtained his BComm and MComm in Economics at the University of Stellenbosch, and completed his doctorate at the University of Cape Town. He spent 8 years working as an economist at the Bureau for Economic Research, University of Stellenbosch and then established a private consulting firm, becoming Director of Medium-Term Forecasting Associates, a company specializing in research into building cycles and building costs. These forecasts are made available to building materials suppliers, building contractors, quantity surveyors and other building professionals. In addition, in 2007 he was appointed Adjunct Professor in the Department of Construction Economics and Management at the University of Cape Town.

Stephen Gruneberg is Senior Lecturer in Construction Economics at the University of Westminster having previously been at University College London and the University of Reading. He has worked on a number of projects, including an international comparison of construction industry productivity, urban regeneration, site management training and consortia in construction. He has written a number of books on construction economics, including Construction Economics: An Introduction; the Economics of the Modern Construction Sector and the Economics of the Modern Construction Firm. His research interests include global construction markets, economic theory applied to the construction production process and specialist firms and markets within the construction industry.

Christian Brockmann holds degrees in civil engineering, construction management, business administration, and economics and teaches in these areas at the University of Applied Sciences Bremen, Germany. His interests in the economic aspects of the construction industry focus on how structures define behaviour, performance and outcomes. During his 16 years as a professional in the construction industry, he became fully aware that mainstream economic theory cannot be applied directly to this field as the structural differences are too important. He has taught as visiting professor at the Asian Institute of Technology in Thailand and at Stanford University in the US.

Göran Runeson, recently retired, is Adjunct Professor at University of Technology, Sydney. After a few years in the Swedish merchant navy and construction industry, he obtained a 1st in Economics and worked as a lecturer in Economics and later Construction Management. He is the author of three books, 160 papers and supervisor of 20 past and present PhD students.

Gerard de Valence is a Senior Lecturer in the School of the Built Environment in the Faculty of Design Architecture and Building at the University of Technology Sydney, where he is Director of the Postgraduate Property Programs. Prior to becoming an academic he had 10 years' experience as an analyst and economist in the private sector, doing research on the property, building and construction industries. He was co-editor of the three volume Building in Value series: *Pre-design Issues, Design and Construction* and *Workplace Strategies and Facilities Management*. He is International Coordinator of the CIB Working Commission W55 on Building Economics.

M. Talat Birgonul is Professor and currently the Director of the Construction Management and Engineering Division in the Civil Engineering Department at Middle East Technical University, Turkey. His primary research interests include engineering economy, health and safety, construction planning, macroeconomic aspects of the construction industry and claim management. Apart from his academic activities, he acts as an expert witness in Turkish courts and arbitral tribunals and gives claim management consultancy service to leading construction companies.

Irem Dikmen is Associate Professor in the Construction Management and Engineering Division of the Civil Engineering Department at Middle East Technical University. Her primary research interests include risk management, knowledge management, strategic management and IT applications in construction. In addition to her research activities, she offers continuing education seminars and consultancy services to construction professionals about international business development and construction project risk management.

Beliz Ozorhon is a Researcher in the Construction Management and Engineering Division of the Civil Engineering Department at Middle East Technical University. Her research initiatives focus on international joint ventures, performance management, decision making and knowledge management.

Jo P. Soeter is Associate Professor of Building Economics at Delft University of Technology. He teaches Building Economics on the BSc Architecture and Urbanism programme and to Master students of Real Estate & Housing and Civil Engineering Management. The special focus of his teaching is on feasibility studies and on individual and social demand–supply relations, with regard to real estate and construction markets. His research activities and publications are concentrated on the development and the adjustment of the Dutch building stock (1960–2020). This is in combination with the changing functional–economic structure and spatial spread of the stock.

Philip W. Koppels is a Researcher in the Department of Real Estate and Housing, Delft University of Technology. His research is focused on the office market. Specifically, on how building and location attributes of offices are related with the economic performance in terms of value in use, rent revenues, capital value, market value and return on investments. More knowledge about this relationship provides for an improved appreciation of future development and design opportunities. Besides his research efforts, he teaches in Real Estate Economics, in relation with urban economics and land use planning.

Peter de Jong is a Lecturer in Design & Construction Management and Economics at Delft University of Technology. He teaches various master and bachelor modules about cost planning in relation with building quality, and the consequences this has on the design, construction and management of buildings. The subject of his doctoral research is High Rise Cost

Modelling. The main focus is on building costs in the conceptual and design phase and their social and economic acceptance and feasibility.

Arturas Kaklauskas holds a Chair in Construction Economics and Real Estate Management and is Vice-Director of the Institute of Internet and Intelligent Technologies at Vilnius Gediminas Technical University. He is a Lithuanian Science Prize Laureate, Expert Member of the Lithuanian Academy of Sciences and Leader of the CIB Study Group SG1 (The Application of Internet Technologies in Building Economics) He is editor of the *International Journal of Strategic Property Management*, the *Journal of Civil Engineering and Management* and *Facilities* for Central and Eastern Europe. He has participated in several European Framework projects and is author of over 200 research publications and seven monographs.

Edmundas Zavadskas is a Professor and past Rector of Vilnius Gediminas Technical University (VGTU) and is currently the First Vice-Rector of VGTU and Head of Department of Construction Technology and Management. He has participated in several European Framework projects. His current research interests are internet based and e-business systems (property, construction and export), decision-making theory and decision support systems. He is the author of over 500 research publications and 14 monographs.

Introduction and overview

Les Ruddock

There is no doubting the importance of the role played by the construction and property industries as being crucial to the development of the economy of any country. Worldwide, these industries employ more than a hundred million people and produce the environment in which we all live and work. The modern context of a rapidly changing society and economy leads to a need to impose a logical view of the economic factors at work in a constantly developing built environment.

The chapters of this book represent an attempt to come to grips with one of the most important areas of research in the built environment today. Studies of economic aspects of the built environment have made great advances in the last 20 years but, in the present state of development of built environment economics, there is certainly no integrating methodology/framework/theory, which neatly fits everything together to form a complete picture. We must acknowledge that the study of the economics of the built environment has to involve a variety of approaches and the application of various techniques. This could be considered to make for a fragmented way of working and, whilst it may not constitute a global approach to a study of the construction and property industries in all its aspects, the various elements can be combined to give a better understanding of the modern built environment.

The contributors of the various components of this text (all respected researchers in the field of construction and property economics) were given the brief of writing on their particular area of expertise and showing state-of-the-art thinking in that field. Each author has taken a particular topic and presented a contribution, which illuminates the present state of knowledge. There is a logical structure to the contents of the book and this prelude to the contributors' chapters puts into context and synthesizes the various issues dealt with and the approaches taken by the writers.

What are these issues, which allow a meaningful use of the expression '*modern* built environment' in the title of this text? The following overview of the contents of this book provides the answer.

Technology transforming built environments

Throughout the course of the twentieth century, the development of new technologies (in the form of electricity, elevators, steel frame construction, curtain walls, for example) led to buildings as we know them today. In the twenty-first century, technology is again transforming how we design and construct buildings and the building process itself, how we operate and maintain the buildings, as well as how their occupants experience and use them. The power of technology to transform economies and societies has been a fundamental part of our history, and advances add value to real estate by creating environments that have liberated human activities from site and climate, intensified space use and facilitated urban development (O'Donnell and Wagener, 2007).

In the opening chapter on the 'Changing Nature of the Built Environment', **Bröchner** takes the view that, whilst many of the large-scale changes in the built environment during the twentieth century occurred in response to changes in transportation technologies, the present force of change is mostly ICT driven. After the defining years of the early 1960s, there have been few or no technical innovations in construction technology or in transportation systems but growing opportunities for raising the efficiency of how consumers and producers use the built environment due to applications of ICT advances.

An interdisciplinary approach to the economics of the built environment

Bröchner sets the scene by explaining that the dynamic of how the modern built environment relates to the economy as a whole, constitutes its changing nature, and he draws on contributions from other disciplines (notably economic geography) to elicit an overview of the forces that have helped to configure the built environment. He explores spatial economics and provides a comprehensive literature review to explain the dynamics of the built environment and its relationship to the local economy, making the point that it is unusual for urban economists to take much interest in the built environment in cities (including such features as the durability of housing) when they study urban growth.

Whilst economic theories can only throw a partial light on architectural changes on the urban scale, the development of spatial economics in recent years shows promising signs of taking the physical nature and configuration of the built environment more seriously but the lack of access to large amounts of data collected and classified according to relevant categories remains a barrier between the highly aggregated urban modelling and smaller case studies. This is the main reason why we are unable to analyse how the built environment is related to innovation in the economy.

By providing an overview of the institutional context[1] within which market forces operate, he shows how rapid changes in economic systems, such as the transition to market economies in Eastern Europe and the reforms introduced in China are associated with transformation of the built environment.[2]

Bröchner considers the economics of asymmetrical information. The study of signalling in markets where there is asymmetric information has been developed in recent years and he poses the questions: Can it be applied to buildings and, more specifically, can architecture be understood as a special form of advertising that both reveals hidden qualities and contains an element of hype? Considering the distinction between informative aspects of advertising and prestige and image effects, he points out examples, such as the Sydney Opera House leading to a new generation of landmark buildings as a means of signalling the inate qualities of buildings. He appraises studies of signalling at both the city and building levels and concludes that historicism and eclecticism are easily analysed.

Development of ICT affects the built environment in many ways. Concerning the physical consequences of new technology on buildings, as well as conspicuous symptoms (such as antenna masts and satellite dishes on buildings) he looks at indirect effects related to types of building, the design process etc.

Musing over whether the revolution in office work brought about by modern ICT, the question is posed: Do obsolescence related factors affect rental value or does a more general law of obsolescence generate rental patterns?

ICT is seldom a full substitute for older technologies or practices but the effects of ICT (both direct and indirect) makes the use of traditional physical resources more efficient.

Approaches to the role of the construction sector in the economy

The relationship between the construction sector and the overall economy is considered in an international comparative study by Lewis, with an examination of the issues associated with the construction sector and the difficulties concerning measurement of the sector based, partly, on the inadequacies of the narrow view of the United Nations definition of the sector, which have led to its neglect by economists. The chapter provides a review of previous studies (empirical work) on the relationship between the construction sector and the level of economic activity in the rest of the economy. Relevant literature goes back to the 1960s, particularly Turin's work on the pattern of construction activity in terms of an economy's stage of development, and Bon's empirical work since the 1980s, in which he identified the inverted U-shaped relationship. Focusing on the link between the construction sector and the economy as a whole, Lewis poses the vital question regarding a causal relationship between the construction sector and the economy: Which way is it?

In the early stages of development, construction can account for a large part of economic activity in the rest of an economy. As an economy develops, construction assumes a relatively less important role. He validates these relationships using time series data for a large number of countries in both the early and later stages of development. Using United Nations data, the relationship is tested and the results validate the view that, in developed countries, the construction industry is relatively small and declining in importance.

Because of its role as a provider of capital assets, the relationship between construction and a country's gross fixed capital formation is an important measure of its contribution to the economy. Whilst there is a steady level in developed countries, this is more divergent for developing countries due to the importance of government expenditure. Construction plays a large role in modernization, with a large proportion going on infrastructure. Infrastructure accounts for 60 per cent of total capital formation but falls to about 25–30 per cent for developed countries. However, the significance of shifts as developed industrialized countries face the replacement of a rapidly decaying installed infrastructure, has started to become an important factor in recent decades.

In another approach to gaining insights into the workings of the economy and its relationship with the construction sector, **Gregori** provides a review of input–output techniques and their application to the construction sector. Starting with the origins of the concept, he explains the development of input–output techniques from Cantillon and Quesnay in the eighteenth century to the ground-breaking work in the 1930s and 1940s, when Leontief introduced technical production relationships in the form of mathematical equations.

The demand (use) and supply matrices from input–output tables provide the analytical framework for economic modelling, and measures based on input–output analysis have been used to quantify the importance of each industry sector, such as construction, in the economy. Calculated indices can be used to show construction's role in the macroeconomic system but do not describe its technology in terms of disaggregated inputs and deliveries to other sectors. The fact that technological structural relationships should ideally be measured in physical units, whilst most tables report flows in volume terms, is an issue but the usefulness of input–output tables for construction is mitigated because construction value-added is relatively small (and intermediate use is large) compared to other sectors.

The question is considered of whether the inverted U-shaped Bon curve holds for volume of construction activity as well as value. As identified by Lewis, at early stages of development, building activity is, at first, growing at an increasing rate and eventually at a decreasing one. As labour and value-added move to services, the construction share declines. Input–output tables can provide useful insights as to whether the construction sector is a growth engine in developed countries as well, as it can generate further output due to intersectoral linkages.

Ruddock and Ruddock put forward the view that highlighting the importance of construction to an economy is a key factor in ensuring that the sector has a high priority for governments' economic agendas. This may be particularly important for developing countries as it can provide a basis for driving the whole development agenda. In an evaluation of the construction sector, a distinction can be made between the value *in* the economy and the value *to* the economy, both of which are dependent upon the definition of the sector used. In the context of the Revaluing Construction agenda of the *International Council for Research and Innovation in Building and Construction (CIB)*, the case is made for a re-evaluation of the construction sector in order to ascertain its true value to the economy and to society.

The changing role of the construction sector, with a focus on the service rendered by buildings, coupled with developing functions of construction firms in terms of diversification and vertical integration with particular emphasis on supply chain control, calls for a new approach and he proposes a measurement model in the context of a mesoeconomic framework (developed by Carassus) to assess the importance and scope of the construction sector beyond the narrow definition of construction activity. Like Lewis, they reiterate the limitation of existing data sources and discuss the need for improvements to national and international data sources.

Given that the construction industry is a major player in capital formation in the economy, Lopes examines the measurement and determinants of investment in structures and investigates the degree to which structures investment is a strong factor in determining economic growth. Undertaking a historical review of the role of the construction sector in economic growth, he analyses the pattern of development of construction investment in the world economy and appraises the validity of endogenous growth theory that considers that improvement in the quality of structures, rather than the quantity, should be the concern of developing countries.

Regulatory issues

Whilst the general institutional context, within which the construction industry operates, is considered by Bröchner, Briscoe examines the most significant effects that changes in government fiscal, monetary and regulatory policies exert on the sector as governments make adjustments to their fiscal and monetary policies in an attempt to achieve their core macroeconomic objectives. He describes the key macroeconomic objectives (economic growth, full employment, control of inflation, balance of payments stability and protection of the environment) that governments seek to achieve and considers that, when aspects of these policies are changed, there may be significant consequences for the construction sector.

Focusing on each of the three areas of policy in turn:

- Firstly, the aims and scope of fiscal policy are described, with particular reference to those aspects of policy which impact on the construction sector – such as taxation constraining the ability of both households and companies to spend on construction products and services; increased levels of public sector spending directed towards capital projects etc. Large-scale capital expenditure programmes aimed at the creation of employment opportunities typically have a high construction content.

- Then, the objectives of monetary policy are considered – tighter monetary policy, with higher credit costs and more restrictive lending adversely affects both construction companies and their clients. Stable prices benefit the construction sector as they serve to contain tender prices and produce an increased likelihood of clients' budget costs being met.
- The main instruments of regulatory policy – building regulations, environmental directives and planning legislation – constrain companies in the design and implementation of construction projects. Relatively recently, many governments have added protection of the environment to their set of macroeconomic objectives – specific regulatory policies to promote efficiency in the use of energy. Policies to deal with environmental issues are likely to have a strong impact on construction, as buildings are significant generators of carbon and are often implicated in poor waste management and pollution practices.

Building cycles and forecasting

The relationship between the demand for construction activity and price constitutes an important part of the work of building economists and, therefore, it is important to include a chapter on fluctuations in demand levels and how to forecast them. The focus of **Snyman**'s chapter is on time series analysis and leading economic indicators, and it provides an overview of the meaning of business cycles with an explanation that research surrounding leading economic indicators is intertwined with research into business cycles.

Two streams have been developed in the search for useful leading economic indicators. He considers the origins of quantitative leading indicators and explains the validity of short-term forecasting based on the tendency for long-term indicators to lead at building cycle turns. The central idea of the indicator approach is to find a batch of statistical time series that conform well to the business cycle and show a consistent timing pattern as leading, coincident or lagging indicators, but this is not so simple – the indicator approach is quite complex both in theory and in practice. He summarizes the cyclical indicators approach as one in which the composite indices are the key elements in an analytical system designed to signal peaks and troughs in the business cycle.

The quantitative approach is contrasted with the qualitative one; the latter being based on subjective evaluations, made by individuals of variables such as business confidence. Business surveys have gained acceptance and respectability as a means of business cycle analysis, based partly on their ability to give a broader and fuller picture. The chapter contains examples of leading economic indicators in practice – both quantitative and qualitative – and Snyman discusses the need to integrate both to improve business cycle analysis.

Going beyond the use of indicators as an aid to short-term economic forecasting, Snyman looks at the long term, with an analysis of business cycles. In a presentation of a history of the analysis of building cycles, reference is made to the groundbreaking work of Kuznets on economic cycles and its implications on transport and housing investment. Several studies indicate that there is not just one building cycle but a different demand cycle for different construction sectors and Snyman's own study on South Africa indicates that it still exists in the private housing market. His assertion is that, because fluctuations in the level of construction directly affect changes in cost inputs and tender prices, improved forecasting methods of cyclical changes in demand levels could enhance cost planning of construction projects.

Construction markets and the global economy

Internationalization has long been on the construction economics and construction management agendas and, in the context of the global economy and the opening up of the international construction market, **Gruneberg** reviews the state of construction markets worldwide. In a discussion of the diverse factors that influence demand in different parts of the world, his chapter provides an examination of the conditions underpinning national income and considers the main influences determining effective demand for construction services. Variation in the performance of construction industries in different parts of the world is determined by the different factors that influence the demand for construction and the chapter contains an investigation of the economic and social situations, which form the backdrop. These factors affect national income and data on national income are used to predict developments taking place in construction markets.

Comparing a large number of countries using constant dollars and exchange rates, he points out that this type of comparison can be indicative for pointing out differences but regrets the paucity of purchasing power parity data for the construction sector. The significance of the construction sector in economies at different stages of development is again emphasized.

Along with economic growth, comes a demand for improved housing and other infrastructure and more diverse building types. Also, with population growth and urbanization, there is inevitably a growth in construction demand. An examination of demographic, political, environmental, economic and technological changes is based on illustrations from a variety of countries. Size of population and changes in the national income are obviously major factors affecting the construction market and examples are used to show the importance of population movements, urbanization, political stability or uncertainty, the economics of climate change and changes in trading patterns.

In the chapter on 'Global Construction Markets and Contractors', **Brockmann** makes the assertion that 'global players need global markets' and explains how construction markets can be characterized as local, yet still there are companies that do business throughout the world. He distinguishes between markets at different levels (regional, national, international, multinational and global) and focuses on the question: What different marketing strategies do construction companies employ that allow separate levels of market to be defined? Both geographical factors and the skills needed for each market allow distinctions to be made.

The normal context for a global player includes product standardization, worldwide branding and the ability to achieve economies of scale but the basic contention is that global players in construction must employ strategies different from those in manufacturing. By designating construction products as contract goods, as opposed to exchange goods (in the terminology of *New Institutional Economics*), there is an assertion that the path to globalization in the construction industry is a distinctive one.

Runeson's and **de Valence**'s focus is the creation of the 'new construction industry', so different from conventional construction as to constitute a different industry. The drivers of globalization and progress in ICT force large firms to grow even larger and the result is a high technology oligopoly developing out of, but separate from, the traditional industry.

They look at *why* and *how* firms grow. Providing a discourse on what economic theory has to say about why firms grow, there is also an examination of types of growth (through the creation of new capital or acquisition of an existing firm). Growth out of regional or national markets has traditionally been the starting point for multinational firms but now, with a situation in which capital and goods can move freely across national borders, many industries are dominated by a small number of firms that satisfy a global market.

What is the 'new construction industry'? Diseconomies of scale may exist for simple construction but, with the addition of the functions of financier, designer, developer and facilities manager, there is a lot of difference between the role of a contractor in a conventional project and, for example, the private supply of public projects. This can lead to considerable economies of scale. Such change has led to a 'new construction industry' and the potential for a small number of giant firms producing, not a service but a product in a business model that owes more to conventional manufacturing than to traditional building and construction. The new industry is, essentially, a modern global manufacturing oligopoly, whose products are packages (complete buildings, designed, financed, built, maintained and operated) as demanded by their clients.

They conclude that the development of the new industry is facilitated by a changing environment with the most important forces behind the change being globalization of the market for construction and progress in communication technology – plus, internally, attitudes enabling firms to grasp opportunities and have the vision to see where it leads. Examples from Australia are used to illustrate the ongoing process as part of the industry moves towards being part of a small number of giant, global, diversified construction firms.

Birgonul et al. also examine the theoretical background relating to multi-national enterprises. They revisit prevailing theories and consider their validity for emerging economy multinationals, which are involved in international construction.

Internationalization of contracting services has long been on the agenda of the construction management literature. However, theories investigating the pattern and impacts of overseas construction projects are confined to companies originating from advanced industrialized countries only. In their chapter, the focus is on the behaviour of emerging economy multinationals operating in international markets. It is argued that learning from international activities is a key concept that explains the success of global players coming from emerging economies and needs to be explored in more detail. Within this context, a frame-work is proposed to analyse the interrelations between competitive scope, reverse knowledge transfer, dynamic capabilities and competitiveness. Based on a study of the reverse knowledge transfer of Turkish contractors working abroad, they propose some strategies to increase their competitiveness in international markets. Even though, in this case, the initial impetus for internationalization was national comparative advantage, the sustainability of success in international markets can be explained by a resource linkage, leverage and learning framework. Specifically for construction companies, such research should be channelled to a learning theme rather than examinations of the national sources of comparative and home advantages only.

Interdependence between the construction and real estate sectors

The link between real estate, development and the construction sector is ana-lysed by Soeter et al. An analysis of the relationships is based on an examin-ation of the Dutch experience – a case study which is not atypical of many relatively free market economies. Their analysis uses an interdependency model based on the linkage: Space market → Property market and Development → Construction market. In such a model, there is no guarantee, even in relatively free markets, that the supply side of the construction market can meet the demand created by a changing society and economy.

An important finding from their study is that the ICT revolution has not necessarily required new construction to allow technological development of the building stock. The existing stock can be adapted for new ICT facilities. For development and construction companies, the market is switching from expan-sion to renewal of the stock,[3] and they predict that renewal and stabilization of the stock of building will constitute the major part of the market in industrial-ized countries for the foreseeable future.

Kaklauskas and Zavadskas then deal with theories of investment in property and real estate, particularly in the context of information, knowledge and intel-ligent technologies. Their chapter is concerned with the subject of investment in real estate at both the micro and macro levels, distinguishing between and contrasting investment in a building and investment in a built and human environment. Nowadays, a built environment is characterized by the intensive creation and use of information, knowledge and intelligent technological applications and it is commonly agreed that use of these applications will significantly speed up built environment processes, improve the quality of the built environment and the effectiveness of decisions made. They provide an examination of the way in which information systems and technologies are revolutionizing the information gathering and investment procedures and decision making in property markets.

Throughout the text, a common feature of the chapters is that they focus on the integration of models with real-world applications and in many cases are based on case studies. Whilst all the topics and themes are themselves significant in their own right, what is important to understanding the economics of the modern built environment is how the various issues addressed by the different chapter authors can be combined to present a more comprehensive picture.

Notes

1 Some aspects of the institutional context (Regulatory Systems) are dealt with by **Briscoe** in his chapter.
2 **Kaklauskas** (Eastern Europe) also looks at the issue.
3 See **Ruddock** (Chapter 4) for data confirming this situation.

Reference

O'Donnell, K. and Wagener, W. (2007) *Connected Real Estate*. Cisco Connected Series, Torworth Publishing: Kent.

Chapter 1

The changing nature of the built environment

An economic perspective

Jan Bröchner

Introduction

The built environment is the accumulated residual history of mostly man-made environmental change. It receives additions from new construction, it is reinvested in, adapted, refurbished and maintained, or the structures are disused, dilapidated and disappear in time, despite their generally extreme durability compared to other goods in the economy. For our purposes, the dynamics of how the modern built environment relates to the economy as a whole constitutes its changing nature. In this chapter, we shall explore the links between the economy and the built environment, relying on varieties of microeconomics: spatial economics and the economics of asymmetrical information, in particular that of signalling in markets; sometimes, it is useful to draw on contributions from other disciplines, notably economic geography, for broader approaches.

Buildings and the physical infrastructure in general are durable and immobile, more so than other resources in the economy. The relative permanence of most built structures explains their role as vehicles for long term transfer of resources, through changes in ownership or mortgaging. It is no accident that the euro banknotes carry architectural motifs, both as symbols of stability and as elements of a generally recognized cultural language. But the very durability also threatens with increasing physical misfit between inherited facilities and present patterns of consumption and production in a given location. The signals emitted by the built environment are reinterpreted over time and may accelerate decay and demolition; however, in other cases, they may lead to greater efforts for maintenance and renovation.

Simulation models for the forecasting of material flows into the buildings stock, or at least the housing stock, have been developed for Germany (Kohler and Yang, 2007), the Netherlands (Müller, 2006) and Norway (Bergsdal et al., 2007). All these models are based on observed patterns for new construction and demolition, but there are no explicit links to economic mechanisms.

An author coming from the field of cultural studies may see the spread of glass architecture as linked to the home as an interactive media centre and driven by a new culture of transparency (McQuire, 2003). However, a simple economic explanation can be sufficient as an alternative. Changes in the use of building materials can be interpreted as reactions to changes in relative prices over the years. During the 25-year period from 1982, sheet, plate and float glass fell slightly in current dollar prices in the US, while prices of cement, ready-mixed concrete and gypsum boards increased to more than twice their 1982 level, according to US Bureau of Labor statistics. It is thus not only architectural fashion and computational technologies for structural analysis (Addis, 2007) that explain why modern office buildings look the way they do.

Since most changes in the built environment give rise to effects that are external to the market, be they negative or positive, the institutional context within which any market forces operate is important. Central and local government devise legal systems for property ownership and use, such as for tenure security and titling in slum areas (UN-Habitat, 2003: p. 107 seqq.; Field, 2005; Méndez, 2006), they engage in planning (Glaeser et al., 2006; Glaeser and Tobio, 2007; Yin and Sun, 2007), issue building regulations, offer subsidies or tax incentives (Pickerill and Pickard, 2007) and sometimes they impose price or rent controls (Gilderbloom and Ye, 2007). Why many structures in US cities have been abandoned, boarded up and demolished has been explained as owner optimal behaviour under uncertainty and in the presence of the legal forfeiture mechanism (O'Flaherty, 1993).

Refurbishment and changes in use of buildings are often particularly subject to a number of institutional effects. The emergence of vacant and obsolete office space in the UK and many other countries with the recession in the early 1990s soon led to a number of conversions for residential purposes (Heath, 2001). In Toronto, the alternative for owners was often seen as demolition in order to create parking lots, whereas the tax consequences made politicians take the initiative to promote residential conversion. This has fitted in with a growing taste for central city amenities and with congestion problems in urban transport systems. For a British provincial city such as Nottingham, office refurbishment and development in the 1980s was clearly influenced by the availability of central and local government subsidies, as well as by a range of institutional factors in the regulatory process and in the traditional lease contracts (Bryson, 1997).

Rapid changes in economic systems, such as the transition to market economies in Eastern Europe and the reforms successively introduced in China, are associated with transformations of the built environment. This allows an identification of current ties between consumption, production and the built infrastructure: more decentralized choice of goods, services and means of transportation shift the distribution between building types according to purposes and also housing densities. The post-socialist city experiences with quick commercialization of city centres in Eastern Europe are typified by Leipzig: there was an initial rise in shopping malls (Kok, 2007) and enterprise zones, which was followed by residential suburbanization and then by a shift to a higher level of investments in the inner city (Nuissl and Rink, 2005). However, there are differences in trajectories; the former German Democratic Republic, which was incorporated into the Federal Republic, experienced more of a rapid suburban sprawl than Hungary and Poland did (Kotus, 2006). These highly visible effects occurred in spite of overall population decline in many cities (Hall, 2006; Turok and Mykhnenko, 2007). There have been major changes in Chinese cities in recent decades, much as in the transition economies in Eastern Europe, but on a larger scale and clearly affected by stronger economic growth (He et al., 2006).

Spatial economics and the built environment

Any urban area can be interpreted as a particular balance between positive and negative externalities of both consumption and production activities, given that there are transport costs. When analysing the dynamics of the built environment and its relations to a local economy, there are two issues in spatial economics or urban theories that come into focus: agglomeration forces in cities and sprawl. While earlier theories (Anas et al., 1998) assumed monocentric cities and were unable to predict or analyse the move towards polycentricity and phenomena like reverse commuting (people who choose to live in the centre and who work in the periphery), there are recent developments, notably the shift in theory towards a consumption perspective on city centres (Glaeser et al., 2001), which will be discussed here. Neglected questions such as the effects of durability of housing and of housing supply conditions have been approached only lately.

In Marshall's *Principles of Economics* (1920), there are three types of agglomeration effects in cities: knowledge spillovers between firms, thickness of markets for specialized inputs and linkages, both backward (generating greater demand for goods) and forward (variety of goods produced). It is difficult to observe directly how knowledge spills over in the built environment and how spillover is influenced by the geometry of the environment, but location patterns for activities reveal something. Localized knowledge spillover effects as measured by agency location attenuate quickly: counts of neighbours matter strongly up to 500 m and then drop sharply, judging by a survey of advertising agencies in New York City (Henderson, 2007). This distance can be compared to the 150 feet limit (approximately 46 m) for spillovers, as measured by property prices, due to new and rehabilitation residential investment in Cleveland, Ohio (Ding et al., 2000). If the effects are so local, it should be possible to assess the effects of the physical form of the local environment.

More recently it has been argued by Duranton and Puga (2004) that there are three basic mechanisms for generating local increasing returns: sharing, matching and learning. *Sharing* means that small fixed costs paid by producers can be spread across larger quantities as markets grow; *matching* mechanisms refer to how larger markets improve the quality and probability of matching; *learning* is exploiting local size for creation and diffusion of knowledge. It should be possible to interpret these three mechanisms in relation to the built environment, but, as shall be seen at the end of this chapter, despite efforts to understand science parks and so-called innovation environments, little is still known about the links between physical form and innovative activities.

Amenities is a concept that is used to explain why different household income groups locate differently in cities; consumer utility depends on non-housing and housing consumption and also on the amenity level $a(x)$ at distance x to the central business district (Brueckner et al., 1999). Households are faced with commuting cost per mile and housing price p per unit q of housing. If the marginal valuation of amenities rises sharply with income, dynamic effects can be generated. Amenities are divided into natural (such as water access), historical (both largely exogenous) and modern amenities (restaurants, swimming pools, etc.) which are endogenous depending as they do on current income level in the neighbourhood etc. Renovation of a central city's historical amenities can link historical to modern amenities. Maintenance of historical amenities may, with time, explain why historical amenities are partly endogenous in this model. More recently, the idea of starting from the endogeneity of historical amenities has been developed in a dynamic perspective by Yonemoto (2007), again with differences between poor and rich people driving the mechanism. Yonemoto thus assumes that a city's historical amenities, which are considered exogenous today, may have been formed endogenously over time. This approach can be simplified and developed in a two-period model with two locations in a city and two income types. In the absence of historical amenities, locations of the rich and the poor are never reversed in this model, since the poor always locate closer to the centre, for increasing population, income and utility levels of each income type. Assuming that the rich leave some historical amenity behind for the residents in the second period, he is able to show that locations are reversed when the population of the first period is moderate, income disparity between the two types is low and the rich are sensitive to amenity.

In an oft-quoted article that established the concept of a Consumer City within spatial economics, Glaeser et al. (2001) identified what they saw as four critical urban amenities:

1   Rich variety of services and consumer goods.
2   Aesthetics and physical setting (including weather; the authors note that 'We have little evidence on the role of architectural beauty').
3   Good public services.
4   Speed (transport speed).

This typology of urban amenities is reminiscent of the principles for generating 'exuberant diversity' in cities, as stated by Jane Jacobs in 1961. Other kinds of amenities have now been explored in spatial models: environmental

(Wu, 2006) and agricultural or rural amenities explaining the French 'periurban' city (Cavailhès et al., 2004).

Geographers who study the development of productive activities in the inner cities have tried to pinpoint the novelty of mixtures from traditional craft production and informal services, to highly specialized advanced-technology firms (Hutton, 2004). After studying two reference cases, London and Vancouver, Hutton concludes that the 'intimacy of engagement between the working and social worlds, and the marked affinity between firms and creative workers for intensely localised environmental conditions (buildings, streetscapes, spaces), also constitute distinctive features'. In a later discussion, Hutton (2006) brings in experiences from the Telok Ayer district in Singapore, too, and gives more emphasis to particular physical features of the inner city environment; he notes a shift from buildings as control – basically, Hutton identifies physical space and form as well as symbolic constructs. In another rich taxonomy typical of urban geographers, Gospodini (2006) identifies creative islands and edges which constitute signifying epicentres that 'usually introduce a "glocalised" landscape of built heritage and innovative design of buildings and public open spaces'.

The study of sprawl is primarily an American concern (Glaeser and Kahn, 2004; Bruegmann, 2005). It is only recently that good data on urban sprawl have emerged. Using remote sensing data for all US metropolitan areas, Burchfield et al. (2006) have analysed sprawl measured as undeveloped land surrounding dwellings. They found that commercial development had become somewhat more sprawling between 1976 and 1992 while residential sprawl remained roughly unchanged during the same period. The extent of sprawl varies dramatically across metropolitan areas, which can be explained by a number of factors. The reliance of a city on the automobile over public transport has an effect. Otherwise, physical conditions matter, such as aquifers that make it possible to have private wells. The density of the road network has no significant influence on sprawl, according to their analysis. There is no consensus among economists today in identifying adverse effects of suburban sprawl (Glaeser and Kahn, 2004), partly because traffic congestion can be relieved by introducing electronic tolls, as in Singapore (Phang and Toh, 2004). However, an analysis of material flows in the development of the building stock in Germany indicates that neighbourhood roads and the supply infrastructure in low-density areas can be more important than the buildings themselves (Schiller, 2007).

It is unusual for urban economists to take much interest in the built environment in cities when they study urban growth. The simplified monocentric model of a circular city with a single centre and suburbs, surrounded by a rural landscape, is typical of the physical structure assumed in most studies belonging to this tradition. One of few exceptions has been elaborated by Lucas and Rossi-Hansberg (2002), whose model allows equilibrium cities where business and housing can be located anywhere in a circular city, reducing the set of basic assumptions to the single idea that firms balance external benefits from proximity to other producers against costs of more commuting for workers.

A polycentric model of this type can be used to study what happens after the loss of capital structures in particular city areas through terrorist action: one computational example given by Rossi-Hansberg himself (2004) assumes that a terrorist attack increases commuting cost as well as the probability of future attacks; a residential area then appears at the centre of the city, and the agglomeration of firms and capital investment both change.

Furthermore, by introducing the durability of housing, it can be explained why US cities tend to increase population more quickly than the rate of decrease when they decline. One of the mechanisms that make decline persistent is that the supply of cheap housing may attract poor people, and low levels of human capital may then create negative externalities or give lower innovation (Glaeser and Gyourko, 2005). The durability of the housing stock has also been shown to create inertia in the face of sudden rises in transportation costs; there is no rapid adaptation resulting in concentrated spatial patterns when energy prices rise, and this is an argument for using planning to limit urban sprawl (Gusdorf and Hallegatte, 2007). Most large US cities are in locations determined by the prevailing high transport costs in the nineteenth and early twentieth centuries, depending on sea and rail transportation; one reason why they have not declined (more) is the heavy investment in (durable) infrastructure already made (Glaeser and Kohlhase, 2004). Cities today, at least in the US, appear to be there to facilitate contact between people.

Although the typical urban model relied on by spatial economists is assumed to be flat, there is in reality the potential for expanding buildings and the transportation infrastructure upwards or downwards. Lifts in tall buildings, high steel structures, underground and elevated rail transit systems all first emerged in the three decades after 1860. Despite its popularity in older utopian views, air transportation has failed to have an impact on city cores, and the 1977 helicopter accident on top of the then Pan Am building in mid-Manhattan, killing four people and scattering more than a hundred rotor blade fragments over several blocks, illustrated its hazards. On the other hand, air traffic between cities has increased vastly, but here the local negative externality problem is almost only that of noise, in addition to carbon dioxide emissions.

The principle of vertical separation of car traffic on several levels, such as with elevated freeways in US cities, or cars either above or below pedestrian traffic to reduce accident risk, has not been an unqualified success because of negative external effects through noise, pollution and visual interference from multi-level roads; vertical separation from pedestrians may reduce social interaction and also be less effective as accident prevention than assumed originally.

Periodically, high-rise buildings have been in favour and then lost it. Singular high buildings for hotels or offices in Scandinavian capitals came to an early halt around 1960. The partial collapse of the recently built Ronan Point block of flats in May 1968 spelled the end of such developments for a long period in the UK. The huge Tour Montparnasse in Paris, completed in 1973, remained an isolated and criticized example while a cluster of tall buildings went up in

La Défense outside the city centre. Only when the attractive forces generated by Docklands tall office buildings – Canary Wharf – proved themselves, was it feasible to erect similar structures in the City of London (Charney, 2007). In these cases, public policies and planning decisions have been dominant; it is seldom that market mechanisms can be studied, as when Hong Kong consumer preferences for attractive views from residential high-rise buildings have been analysed, indicating, when set against construction costs, that buildings with a sea view could be extremely tall (Chau et al., 2007).

If there is one major shift in preferences among consumers and producers, defining the modern built environment, it takes place in the early 1960s. Three books that influenced public opinion came out in the same year, 1961: Jane Jacobs with her *Death and Life of Great American Cities*, Lewis Mumford with *The City in History* and also Jean Gottman's *Megalopolis*, where a European researcher analysed the Northeastern US seaboard and all of them questioning a sprawl society shaped by automobile transportation. It is the link to change in transportation systems that continues in the following year, which saw the demolition of major examples of railway architectural heritage: the Euston Arch and Euston Station in London went along with Pennsylvania Station in New York. At the same time, there were initiatives that amounted to a revival of walking: Strøget in Copenhagen was pedestrianized in 1962, just as pedestrian suburban shopping malls in the US began to appear (Kostof, 1992). Also, sea transportation and the industrial importance of waterfront location had by now undergone changes: in San Francisco, the Ghirardelli chocolate factory was redeveloped in a waterfront site, and the new mayor of Baltimore declared in 1963 that the Inner Harbor would be his top priority for redevelopment. In many countries, the financial restraints on reinvestment in older buildings were relaxed at this point or even replaced by incentives, as in France with the 1962 *Loi Malraux* offering tax deductions for restoration work. Depending partly on national macroeconomic considerations, there was a time lag before most public policies were adjusted to this change in preferences; thus, the Swedish restrictions on lending to housing refurbishment were not lifted until the early 1970s, when declining intra-European labour immigration had lowered demand for new flats drastically as well as construction employment. By the mid-1970s, the 'large-scale destruction' of British cities began diminishing (Stamp, 2007: p. 10). Ultimately, and in many parts of the world, these changes in preferences were to be mirrored in the now familiar physical doughnut form of development with city core revival, tracts of declining middle and outer city areas, surrounded by extensive and spreading developments beyond city limits into rural areas (Hall, 2006).

After the defining years in the early 1960s, there have been few or, rather, almost no radical technical innovations in construction technology or in transportation systems. Instead, there have been growing opportunities for raising the efficiency of how consumers and producers use the built environment – and transportation – due to applications of ICT advances. Nevertheless, the rate and direction of change has been influenced by the oil crisis and currently by the urgency of reducing greenhouse gas emissions.

Today, it is possible to see that the critical voices of the early 1960s were not without foundation. Using US data for the 1950–1990 period, Baum-Snow (2007) has now found that one new highway passing through a central city reduces its population by about 18 per cent, and that aggregate city population would have grown by about 8 per cent had the interstate highway system not been built.

Nodes or terminals where people or goods change modes of transportation – land/land, sea/land – have always been important for the rise of agglomerations. Rappaport and Sachs (2003) have analysed the US as a coastal nation in this perspective. Just because of the nodal function, the accumulated effects of the global changes in the pattern of where shipyards are located and also of the effects of the shift to containers and in general the role of sea transport have had considerable effects on many cities. There has been a concentration of sea transport and of major ports with ever larger container ships. As in Baltimore, the decline has been followed generally by public policies (Bunce and Desfor, 2007) based on the insight that liberated former harbour areas could be added to the supply of central sites for new construction, not least creating attractive settings for residential development. Seafront and riverside locations are hazardous, however. Slowly rising sea levels, hurricanes and tsunamis pose threats, while rivers may rise seasonally due to heavy snowfall or rain. As the Stern Review (2006) has indicated, many large cities (St Petersburg), the Nile Delta, the Caribbean and other regions are in danger of flooding as a consequence of climate change. Artificial islands in the Arabian Gulf as well as land areas reclaimed from the North Sea are challenging locations for building. On the other hand, proximity to deep cold water is an advantage for reducing electricity needs, because hydrothermal cooling can be used to run air-conditioning systems, as in Toronto and Stockholm, also freeing cooling unit space in existing office towers (The Economist, 2007).

Buildings as signals

The study of signalling in markets where there is asymmetric information has been developed in recent years (Riley, 2001). One stream of investigations concerns why firms engage in advertising that goes beyond the provision of product information. Basically, customers are unable to discern quality or at least less able to do so than producers are. Economic theory applied to advertising has been based on Nelson's (1974) distinction between search qualities and experience (meaning unobservable before purchase) qualities. Ackerberg (2003) makes a distinction between 'informative' aspects of advertising and 'prestige' or 'image' effects, which it is clearly possible to transfer to the realm of architecture. More recently, Johnson and Myatt (2006) have introduced a dichotomy between pure hype advertising (for a mass market) and real information (for a niche market). Hype is what shifts the demand curve outwards and real information often rotates the demand curve, according to their framework. These asymmetries could explain signalling on the building level and on the city level, all affecting how the built environment is transformed over time.

Can architecture be understood as a special form of advertising that both reveals hidden qualities and contains an element of hype? Exteriors of buildings are a source of information about their interior spaces. Both exteriors and interiors provide information on the activities that take place within buildings. The solidity of heavy stone façades at least used to signal financial solidity, while the modern widespread use of glass façades is not only explained by their falling relative prices, as mentioned initially in this chapter, but also by their ability to convey what is going on inside: a media firm can express itself with translucent offices, where the evening scenery with inside activities makes an impression on potential buyers of their services – as well as carrying a message in the labour market, attracting a specific profile of potential employees.

Historicism and eclecticism in architecture are easily analysed in terms of signalling. In recent years, Las Vegas with its 1990s foreign and historicist hotel settings and Shanghai with its invitation in 2000 to international architects, drawing inspiration from Britain, Canada, Germany, Holland, Italy, Spain and Sweden (Denison and Ren, 2006, p. 235; Xue and Zhou, 2007) offer good examples, as well as the Californian townhouses replicated as Beijing upper middle class housing (Wu, 2004). Although not unknown, it is less frequent to see advertising that recycles historical styles or copies foreign models.

For owner-occupied housing in many economies, the transition from traditional construction materials and methods as well as building layouts is, or has been, associated with signalling of household attitudes, lifestyles and social status. Studies of these phenomena have usually been undertaken by sociologists or anthropologists. Thus the implications of shifting from traditional construction to more modern concrete-based houses in an Ecuadorian village have been interpreted as architectural conspicuous consumption, partly in the light of the economic theory of signalling, by Colloredo-Mansfeld (1994), a social anthropologist. Adoption of *feng shui* principles for design of buildings, far away from their geomancer origins, probably belongs in this context. It should also be possible to understand some of the strength of the current trend of 'green building' in many advanced economies in a similar way.

The true long-term environmental qualities of a building are difficult to assess for potential buyers. There could, therefore, be an opposing force in the market, reducing investment in higher environmental qualities because of the seller/buyer asymmetry of information. This can be mitigated by the introduction and application of third-party certification schemes for buildings (Lützkendorf and Speer, 2005). Schemes such as the UK BREEAM environmental assessment method and the more recent US LEED Green Building Rating Systems both reduce and consolidate long-term quality information into four levels, although differently defined (Cole, 2006). Just as rating systems for property assets relied on by banks may provide incentives (Lützkendorf and Lorenz, 2007), these schemes should reduce the information asymmetry by lowering buyer search costs for technical information.

Furthermore, there is the ambiguous relation between signalling and building maintenance. For a seller of durable assets, investing in pre-transaction asset improvement and maintenance can be seen as a mechanism for signalling better quality (Ben-Shahar, 2004). Such an investment, again explained by asymmetric information in the market, might constitute a waste of resources. A focus on this mechanism is in stark contrast with earlier literature on urban renewal, where it was usual to refer to simple game theory – the prisoner's dilemma – in order to predict neighbourhood decline because of negative external effects arising from a few individual landlords' faulty maintenance of their properties (Davis and Whinston, 1961). Ceteris paribus, renovation is more likely and more extensive and/or expensive for older buildings, according to Helms (2003), who has studied gentrification in the Chicago area and the determinants of urban housing renovation. Also, renovators prefer low-density buildings with large living spaces. A building in a historic neighbourhood is more likely to be renovated, but the level of renovation expenditure is no different from the level to be expected for a similar building in a newer neighbourhood.

Architecture differs from advertising in the durability of the medium. Over a long stretch of time, signals emitted by buildings are subject to human inter-pretation and subsequent reinterpretation which may hasten decay because the interpretation affects the will to maintain them, to reinvest in them and can also lead to demolition regardless of technical status. Over time, the original message, its value as a clue to hidden qualities and its immediate contribution to consumer utility functions may change.

There have been many attempts to find patterns in economic obsolescence of buildings without explicitly accounting for the signal issues. A study by Bradley and Kohler (2007) of the building stock in Ettlingen, a medium-sized German town, confirms what Meikle and Connaughton (1994) found for English hous-ing, namely that older houses tend to survive longer than more recent stock. Office rental data from Kuala Lumpur in 1990 were regressed on a range of building quality variables by Khalid (1994), who found that appearance, flexi-bility and quality of building engineering services have the main impact; this impact was significant only in the first 15 years after completion of the building. That failures in buildings due to deterioration are far from being a prime cause of housing refurbishment emerges from the Finnish study reported by Aikivuori (1996). In the City of London office market, Barras and Clark (1996) using data from the 1980s and 1990s found that buildings older than 30 years showed little age-related depreciation, and Baum and McElhinney (1997), comparing London office property data for 1986 and 1996 found that obsolescence-related factors (such as configuration, internal specification, external appearance) and physical deterioration were a better explanation of rental value than simple age. In 1996, 'high specification buildings, particularly in services' were more important than configuration, which had been the dominant factor in 1986. Now, we cannot be sure whether these findings have to do with a single revolution in office work brought about by modern ICT – or whether there is a more permanent law of obsolescence that generates these patterns. Two recent US studies with hedonic models, where price is a function of property attributes and the time of sale, distinguish between functional obsolescence (due to technological change), physical condition and location quality. For retail properties, there was a critical point after the first 16 years, when the rate of change for functional obsolescence dropped markedly (Colwell and Ramsland, 2003). But when Corgel (2007) analysed hotel sales, the breakpoint was found to occur almost 30 years after construction; after 28 years, there was even a positive relation between age and sale price (0.7 per cent per year). It is tempting to identify this effect as a consequence of signalling robust qualities or, to put it differently, slow branding of the building. There is also a question of what minor alterations and refurbishment contribute.

Built structures that are highly visible and easily recognized symbols of a particular economic and political system are especially prone to sudden demolition: it is hardly necessary to mention the Bastille, the Berlin Wall and the New York World Trade Center. A more gradual shift was evident when the 1972 demolition of the 1950s public housing Pruitt-Igoe project in St. Louis was followed by the disappearance of similar structures in many other countries. The Ronan Point partial collapse in 1968, already referred to, eventually led to full demolition of the building in 1986. To some extent, the current wave of rebuilding city areas that were transformed by post-war reconstruction in the UK (Thurley, 2007) and other European countries is due to their connotations.

Landmarks are signals. The Sydney Opera House project, initiated in the late 1950s but finished only in 1973, has led to a new generation of landmark buildings, encouraged or invested in by city governments in many countries. These buildings are obvious means of signalling innate qualities of cities, devised to attract temporary visitors or more permanent settling of firms and individuals. Often, they are intended to symbolize urban revival; the major shift in employment in many cities from the manufacturing sector to services, in particular the growth of advanced business services, is probably the best explanation why urban managerialism turned into urban entrepreneurialism (Harvey, 1989) and of the phenomenon of city branding relying on celebrity architects. Estimates of the additional tax revenue for Bilbao related to millions of extra visitors because of its Guggenheim museum opened in 1997 are impressive (Jencks, 2005: p.18). Temporary events, however, and Olympic Games in particular, are more uncertain in their effects on the local economy. The Guangzhou changes to the built environment beginning in 1998, in a country where numerous cities, not only in the Pearl River Delta, attempt to develop themselves into 'international metropolitan cities', and rely on landmark buildings, has been analysed by Xu and Yeh (2005). These landmarks, or iconic buildings to use Jencks' (2005) terminology, often stretch the limits of technological innovation, sometimes dissolving the sense of the solidity of the built environment. Again, they may blend a historicist surface with modern technology for the underlying structure; or as an alternative they affect a stage-prop surface of extreme innovation, hiding the mundane technology used for the load-carrying structure. There is often a paradox in relying on patterns from microscopic biomimetics for designing vast innovative structures that appear to defy the laws of nature. Much of this nature signalling would have been impossible to achieve without the computational possibilities developed in recent years.

Indoor comfort and global climate

In a global perspective, there are two opposing trends in the residential building stock that concern energy use and greenhouse gas emissions; the first is the movement away from traditional buildings and traditional construction methods in many regions, and the second and much weaker is in many ways a retreat to buildings that store energy, are ventilated and are heated or cooled in a modern version of old principles, so-called passive houses. Throughout, there is a considerable increase in electricity consumption among households, in many regions for purposes of air conditioning, but also because of more use of a wide range of small appliances, and the consequent upstream increase of greenhouse gas emissions caused by electricity supply conditions (International Energy Agency, 2007). Economic growth brings with it a convergence of life-styles, apparently also leading to an international convergence of indoor climate preferences, and growth has come to imply more air conditioning (Brager and de Dear, 2003; Addis, 2007). The fulfilment of heating or cooling needs or both, depending on the climate zone, has consequences for the built environment, and not only for buildings, as when supply systems for renewable energy transform the surrounding landscape – wind turbines on high pylons, large areas of photovoltaic elements and biofuel agriculture. Efforts to improve energy storage capacity in buildings and to improve their insulation may also produce visible consequences. Seemingly slight changes in our abilities to store heat or provide low temperatures in buildings may influence both transportation and the entire urban structure: home freezers happened to relocate food storage in urban areas and constitute one of the forces behind big shopping malls in the periphery, thus contributing to sprawl.

What would be the consequences for traditional buildings if there is a major rise in energy prices? It seems that there is a minimum technical standard level that buildings must attain before they attract reinvestment; below this level, demolition or abandonment are the alternative courses of action. In particular, structures must be able to provide a modern indoor climate after refurbishment. This probably contributes to phenomena such as the disappearance of almost all low-rise housing and old alleys in central parts of major Chinese cities and their replacement by high-rise buildings (Denison and Ren, 2006).

A rise in energy prices or taxes affects the built environment in two ways: thermal comfort costs and transportation costs both rise. However, changes in building regulations, information campaigns and subsidies complicate the analysis of consequences for new or existing buildings. The situation is obviously different in different climate zones and depends on income levels. Swedish data indicate that rising energy prices led to improvements in energy efficiency between 1972 and 1985, but that the subsequent development in both the existing stock of buildings and in new construction levelled off with almost no change during the 1990s (Nässén and Holmberg, 2005). Since 1995, newly constructed blocks of flats have increased energy use per floor area unit. A series of design standards has been introduced in China since the 1980s, intended to raise energy efficiency in new and existing buildings, but survey respondents among property developers report that about one fifth of new buildings shows the ambitious degree of energy efficiency required of new public buildings (Liang et al., 2007). Across the globe, it seems that environmental sustainability projects are still few and far between; they can often be understood as having signal functions as outlined earlier in this chapter.

Information and telecommunication technologies

The development of information and telecommunication technologies (ICT) affects the built environment in many ways. There are conspicuous symptoms such as mobile antenna masts and satellite dish antennas on dwellings, which are direct consequences of new technologies. However, there are several indirect effects, on the urban scale, related to the demand for particular types of build ings, on the design process, and derived from various forms of embedded ICT. It appears that ICT is seldom a full substitute for older technologies or practices; broadly speaking, it makes the use of traditional physical resources more efficient.

While the development of transportation technologies in the nineteenth century led to gashes in the urban structure, as when railways were brought into Victorian cities (Kellett, 1969), ICT is far less disruptive. The complexity of cities is increasing along with the number and type of interactions (Moss and Townsend, 2000). With 'ubiquitous computing' and preferences for living in rural areas, more sprawl could be expected. However, telecommunications are less than a perfect replacement for face-to-face contacts. Thus, Gaspar and Glaeser (1998) found it difficult to expect that telecommunications would have major impacts on city growth. The Internet is unlikely to change the needs for physical proximity in trade: there are codifiable and uncodifiable messages, and there is the dominance of uncodifiable messages; the dominance of the latter explains why physical proximity and face-to-face contacts remain important for the economic geography of production (Leamer and Storper, 2001). At least in France, it seems that ICT is not much of a substitute for face-to-face communication in cities; rather, it is a complement (Charlot and Duranton, 2006).

There are several studies of how ICT-related firms choose to locate in cities, such as one of multimedia firms in Sydney (Searle and De Valence, 2005). Tendencies to clustering can be explained by the need for face-to-face contacts, knowledge spillovers of other types, and perhaps the clustering forces were stronger in the mid-1990s than today because of more geographically concentrated access to advanced ICT in certain parts of cities at that time.

Does telecommuting affect household relocation? Not much, it seems. There is diversity of telecommuting and physical commuting within households, which influences location choice, and the stage of the household lifecycle is important, just as income and education are, according to an investigation of telecommuting in the Netherlands (Muhammad et al., 2007).

Going back not many decades, the physical consequences for office buildings of mainframe computerisation were obvious. Raised floors were needed for cabling, and the technical solutions required for cooling appeared to make many office buildings obsolete. For the financial sector, the London Big Bang in 1986 with its deregulation and change of the Stock Exchange system together with the character of ICT support gave rise to a need for large dealing floors and other types of new space. Such changes are part of the background to ground-scrapers – large, low-rise, deep-plan office buildings – in London (Carmona and Freeman, 2005). Although productivity of office work in general rises and there is redistribution between front office and back office activities, the separation appears to be of little consequence (Dixon, 2005). Wireless broadband networks help to explain why the 'office of the future' can be implemented in a range of settings. Facilities related to manufacturing and distribution of goods are another matter; the effect of ICT on supply chains for goods is reduction and relocation of goods storage, thus a new structure of demand emerges for types, sizes and location of warehousing (Thompson, 2005). Perhaps the greatest impact of current ICT is that the existing built environment can be used, adapted and reused more efficiently. The shift from a manufacturing to a service economy is mirrored by call centres accommodated in old factories.

Markets for properties and space in buildings should reach a higher level of efficiency with web portals that improve the overview of possible alternatives and support a more precise matching of demand and supply. It is a general observation that broker roles are transformed by ICT, and a study of the US residential real estate industry by Sawyer et al. (2005) confirms this. An indication that there are efficiency effects is given by study of residential transactions in 1999 from the North Texas Multiple Listing Service (Ford et al., 2005), where it was found that properties listed on the Internet took slightly longer to sell and sold for slightly higher prices.

ICT tools have had many consequences for design activities (Kalay, 2006). New architectural solutions are found. The effect of new modelling and simulation processes in engineering on problem-solving takes place in three major areas: judgment and validation, creating new combinations of technologies and components, and 'design conversations', according to Dodgson et al. (2007).

The need for physical prototypes is reduced, a new generation of software tools allows designers and analysts to 'co-simulate' across different simulation tools rather than solving problems piecewise. The ability to visualize results as a base for knowledge exchange between experts has been increased. These developments have been crucial for many landmark and iconic buildings where the signal effect relies on innovative structures and uses of materials. Integration between design and actual construction is facilitated through building information models, which is one of the reasons why prefabrication has gained in importance; there is higher precision in both the production of components and in site assembly operations.

In many ways, new cars are descendants of the Ford Model T; much of the difference lies in embedded ICT. Although the level of control technology is not the same as for cars, modern buildings and the built infrastructure are increasingly filled with control technology that raises their efficiency in use. In a sense, buildings are dematerialized when ICT helps achieving a desired performance with less physical input. Sensor technology, wireless or not, lies behind the concept of intelligent buildings (Glover et al., 2004). This development also concerns existing buildings, where lifts may be subject to remote monitoring, climate control can be improved in ways that raise energy efficiency, and new electronic access systems installed. An important psychological issue that arises is that of combining sophisticated systems with occupant possibilities of control (Leaman, 2003; Barlow and Fiala, 2007). There are climate zones where better ICT support for charging residents according to individual metering of heating can make an important contribution to energy efficiency.

While ICT seems insufficient as a full substitute for physical protection from crime, it does have effects in the built environment, not only in a surveillance city such as London with high CCTV camera density. It has been claimed that the emergence of US type suburban sprawl in Mediterranean countries is partly caused by alarm systems that reduce the traditional risk of theft in isolated locations (Munoz, 2003). Nevertheless, there is a market for physically gated communities in many countries (UN-Habitat, 2006: pp.146–147; Vesselinov et al., 2007). As a parallel, despite the development of electronic warfare, there are the mock-up cities used for training soldiers to perform military operations in a hostile urban environment.

Electronic systems for road charges (Phang and Toh, 2004) and parking charges (Ignaccolo et al., 2006) are used to relieve congestion and thus raise road capacity. Another example of ICT applications that raises efficiency in infrastructure use is when vehicles activate red–green signals by passing over induction cables embedded in the road surface. Having cars equipped with GPS that supports orientation further increases the efficiency of private transportation. Similarly, the ICT support for coordinating airplane movements raises capacity and reduces the need for new construction. Runway capacity at congested airports has been increased greatly through innovative co-operative procedures (Tether and Metcalfe, 2003).

Through sensor technology, the monitoring and maintenance of structures, such as bridges, can depend less on periodical, planned repair and replacement; instead, even periodical inspection is being phased out in favour of more continuous status monitoring relying on ICT for remote control (Bergmeister and Santa, 2001). The use of remote monitoring allows building less costly bridges, while it also leads to an increased reliance on remedial action, including carbon fibre tape reinforcement (Casas et al., 2002).

Since ICT leads to technical integration along the time axis, from the design phase through construction and into the facilities management phase, one cannot exclude consequences for the organization of industry. Design specialists in building services might find it profitable to integrate into performance contracts over several years for the facility in use (Bröchner, 2003). However, the ICT also supports routinization and standardization of facilities services, management and internal training, which appear to contribute to the emerging large internationally active facilities management groups (Bröchner, 2008).

Supporting innovation

Economic growth in any country used to have an obvious link to investment in buildings and other fixed assets. Today, growth in advanced economies is understood as dependent on innovation. Innovation activity, measured by patents, has been found to be related to urban population density (Andersson et al., 2005), but there is a remaining gap between analysis at the regional level (Acs, 2002) and the local physical structures. On the other hand, there are case studies of small-scale, laboratory-style, teamwork-enhancing 'innovation environments' designed for that very purpose (Kristensen, 2004; Haner, 2005; Van der Lugt et al., 2007), but with little or no measurement of actual performance. Science parks and business incubators have physical aspects as built environments, but the effects are difficult to identify and isolate. However, a study of six university research centres in the US does reveal connections between spatial features recorded, based on space syntax (Hillier and Hanson, 1984) and innovation process variables such as publications and patents (Toker and Gray, 2008). The link between workspace features and innovation outcomes appears to be the unscheduled face-to-face contacts between scientists.

The Schumpeterian concept of creative destruction is seldom invoked for the stock of buildings, except ironically as when the use of the term 'obsolescence' in US urban redevelopment projects is called a 'neoliberal alibi for creative destruction' (Weber, 2002). Whereas the built form of prisons and other early nineteenth century buildings is easily translated into social control (Markus, 1993), it seems far more difficult to design an environment that evidently supports innovatory activity. Perhaps it is just a matter of minimum satisfaction of ordinary building user needs, relying on the accumulated knowledge from user studies (Leaman, 2003).

Concluding remarks

One of the messages in this chapter is that while many of the large-scale changes in the built environment during the twentieth century occurred in response to changes in transportation technologies, the present force of change is mostly ICT. The visible consequences of this shift are difficult to detect because its strongest effect is that of adaptive reuse of the built environment and the rising volume of embedded ICT in buildings and their infrastructure. These developments allow for changes in employment patterns and household lifestyles, and also support a higher level of energy efficiency in attaining a better indoor climate.

Economic theories, however, are far from being able to throw more than a partial light on architectural changes on the urban scale such as those mapped by Kostof (1992). The development of spatial economics in recent years shows promising signs of taking the physical nature and configuration of the built environment more seriously. But the lack of access to large amounts of data collected and classified according to relevant categories remains a barrier between the highly aggregated urban modelling and smaller case studies. This is the main reason why we are unable to analyse how the built environment is related to innovation in the economy. Better data would also make it possible to transform and adopt concepts that related disciplines use for understanding the sign nature of buildings, a development that would throw new light on what is called obsolescence. To take an obvious application, our ability to predict effects of climate policies on decentralized decisions by households and other property owners and users would be strengthened.

References

Acs, Z.J. (2002) *Innovation and the Growth of Cities*. Edward Elgar: Cheltenham

Ackerberg, D.A. (2003) Advertising, learning, and consumer choice in experience goods markets: an empirical examination. *International Economic Review*, 44(3), 1007–1040.

Addis, B. (2007) *Building: 3000 Years of Design Engineering and Construction*. Phaidon: London.

Aikivuori, A. (1996) Periods and demand for private sector housing refurbishment. *Construction Management and Economics*, 14(1), 3–12.

Anas, A., Arnott, R. and Small, K.A. (1998) Urban spatial structure. *Journal of Economic Literature*, 36(3), 1426–1464.

Andersson, R., Quigley, J.M. and Wilhelmsson, M. (2005) Agglomeration and the spatial distribution of creativity. *Papers in Regional Science*, 48(3), 445–464.

Barlow, S. and Fiala, D. (2007) Occupant comfort in UK offices – How adaptive comfort theories might influence future low energy office refurbishment strategies. *Energy and Buildings*, 39(7), 837–846.

Barras, R. and Clark, P. (1996) Obsolescence and performance in the Central London office market. *Journal of Property Valuation and Investment*, 14(4), 63–78.

Baum, A. and McElhinney, A. (1997) *The Causes and Effects of Depreciation in Office*

Buildings: a Ten Year Update. Department of Land Management and Development, University of Reading.

Baum-Snow, N. (2007) Did highways cause suburbanization? *Quarterly Journal of Economics*, **122**(2), 775–805.

Ben-Shahar, D. (2004) Productive signaling equilibria and over-maintenance: an application to real estate markets. *Journal of Real Estate Finance and Economics*, **28**(2/3), 255–271.

Bergmeister, K. and Santa, U. (2001) Global monitoring concepts for bridges. *Structural Concrete*, **2**(1), 29–39.

Bergsdal, H., Brattebø, H., Bohne, R.A. and Müller, D.B. (2007) Dynamic material flow analysis for Norway's dwelling stock. *Building Research and Information*, **35**(5), 557–570.

Bradley, P.E. and Kohler, N. (2007) Methodology for the survival analysis of urban building stocks. *Building Research and Information*, **35**(5), 529–542.

Brager, G.S. and de Dear, R.J. (2003) Historical and cultural influences on comfort expectations. In: Cole, R.J. and Lorch, R. (eds) *Buildings, Culture and Environment: Informing Local and Global Practices*, pp. 177–201. Blackwell: Oxford.

Bröchner, J. (2003) Integrated development of facilities design and services. *Journal of Performance of Constructed Facilities*, **17**(1), 19–23.

Bröchner, J. (2008) Construction contractors integrating into facilities management. *Facilities*, **26**(1/2), 6–15.

Brueckner, J.K., Thisse, J.-F. and Zenou, Y. (1999) Why is central Paris rich and downtown Detroit poor? An amenity-based theory. *European Economic Review*, **43**(1), 91–107.

Bruegmann, R. (2005) *Sprawl: A Compact History.* The University of Chicago Press: Chicago.

Bryson, J.R. (1997) Obsolescence and the process of creative reconstruction. *Urban Studies*, **34**(9), 1439–1458.

Bunce, S. and Desfor, G. (2007) Introduction to 'Political ecologies of urban waterfront transformations'. *Cities*, **24**(4), 251–258.

Burchfield, M., Overman, H.G., Puga, D. and Turner, M.A. (2006) Causes of Sprawl: A Portrait from Space. *Quarterly Journal of Economics*, **121**(2), 587–633.

Carmona, M. and Freeman, J. (2005) The Groundscraper: exploring the contemporary reinterpretation. *Journal of Urban Design*, **10**(3), 309–330.

Casas, J.R., Ramos, G., Diaz-Carrillo, S. and Guemes, J.A. (2002) Intelligent repair of existing concrete structures. *Computer-Aided Civil and Infrastructure Engineering*, **17**(1), 43–52.

Cavailhès, J., Peeters, D., Sékeris, E. and Thisse, J.-F. (2004) The periurban city: why to live between the suburbs and the countryside. *Regional Science and Urban Economics*, **34**(6), pp. 681–703.

Charlot, S. and Duranton, G. (2006) Cities and workplace communication: Some quantitative French evidence. *Urban Studies*, **43**(8), 1365–1394.

Charney, I. (2007) The politics of design: architecture, tall buildings and the skyline of central London. *Area*, **39**(2), 195–205.

Chau, K.-W., Wong, S.K., Yau, Y. and Yeung, A.K.C. (2007) Determining optimal building height. *Urban Studies*, **44**(3), 591–607.

Cole, R.J. (2006) Shared markets: coexisting building environmental assessment methods. *Building Research and Information*, **34**(4), 357–371.

Colloredo-Mansfeld, R. (1994) Architectural conspicuous consumption and economic change in the Andes. *American Anthropologist*, **96**(4), 845–865.

Colwell, P.F. and Ramsland, M.O. (2003) Coping with technological change: The case of retail. *Journal of Real Estate Finance and Economics*, **26**(1), 47–63.

Corgel, J.B. (2007) Technological change as reflected in hotel property prices. *Journal of Real Estate Finance and Economics*, **34**(2), 257–279.

Davis, O.A. and Whinston, A.B. (1961) The economics of urban renewal. *Law and Contemporary Problems*, **26**(1), 105–117.

Denison, E. and Ren, G.Y. (2006) *Building Shanghai: The Story of China's Gateway.* John Wiley & Sons: Chichester.

Ding, C., Simons, R. and Baku, E. (2000) The effect of residential investment on nearby property values: Evidence from Cleveland, Ohio. *Journal of Real Estate Research*, **19**(1/2), 23–48.

Dixon, T. (2005) The impact of information and communications technology on commercial real estate in the new economy. *Journal of Property Investment & Finance*, **23**(6), 480–493.

Dodgson, M., Gann, D.M. and Salter, A. (2007) The impact of modelling and simulation technology on engineering problem solving. *Technology Analysis and Strategic Management*, **19**(4), 471–489.

Duranton, G. and Puga, D. (2004) Micro-foundations of urban agglomeration economies. In: Henderson, V. and Thisse J.-F. (eds) *Handbook of Regional and Urban Economics*, Vol. 4, pp. 2063–2117. North-Holland: Amsterdam.

The Economist (2007) A cool concept. *Technology Quarterly*, June 9, 4–5.

Field, E. (2005) Property rights and investment in urban slums. *Journal of the European Economic Association*, **3**(2–3), 279–290.

Ford, J.S., Rutherford, R.C. and Yavas, A. (2005) The effects of the internet on marketing residential real estate. *Journal of Housing Economics*, **14**(2), 92–108.

Gaspar, J. and Glaeser, E.L. (1998) Information technology and the future of cities. *Journal of Urban Economics*, **43**(1), 136–156.

Gilderbloom, J.I. and Ye, L. (2007) Thirty years of rent control: a survey of New Jersey cities. *Journal of Urban Affairs*, **29**(2), 207–220.

Glaeser, E.L. and Gyourko, J. (2005) Urban decline and durable housing. *Journal of Political Economy*, **113**(2), 345–375.

Glaeser, E.L., Gyourko, J. and Sachs, R.E. (2006) Urban growth and housing supply. *Journal of Economic Geography*, **6**(1), 71–89.

Glaeser, E.L. and Kahn, E.M. (2004) Sprawl and city growth. In: Henderson, V. and Thisse J.-F. (eds) *Handbook of Regional and Urban Economics*, Vol. 4, pp. 2481–2527. North-Holland: Amsterdam.

Glaeser, E.L. and Kohlhase, J.E. (2004) Cities, regions and the decline of transport costs. *Papers in Regional Science*, **83**(1), 197–228.

Glaeser, E.L., Kolko, J. and Saiz, A. (2001) Consumer City. *Journal of Economic Geography*, **1**(1), 27–50.

Glaeser, E.L. and Tobio, K. (2007) *The Rise of the Sunbelt.* Harvard Institute of Economic Research, Discussion Paper No. 2135, April.

Glover, N., Corne, D. and Liu, K. (2004) Information technology, communications and artificial intelligence in intelligent buildings. In: Clements-Croome, D. (ed) *Intelligent Buildings: Design, Management and Operation*, pp. 101–152. Thomas Telford: London.

Gospodini, A. (2006) Portraying, classifying and understanding the emerging landscapes in the post-industrial city. *Cities*, **23**(5), 311–330.

Gottmann, J. (1961) *Megalopolis: the Urbanized Northeastern Seaboard of the United States.* Twentieth Century Fund: New York.

Gusdorf, F. and Hallegatte, S. (2007) Behaviors and housing inertia are key factors in determining the consequences of a shock in transportation costs. *Energy Policy*, **35**(6), 3483–3495.

Hall, P. (2006) Aged industrial countries. In: Oswalt, P. and Rieniets, T. (eds) *Atlas of Shrinking Cities*, p. 144. Hatje Cantz: Ostfildern.

Haner, U.-E. (2005) Spaces for creativity and innovation in two established organizations. *Creativity and Innovation Management*, **14**(3), 288–298.

Harvey, D. (1989) From managerialism to entrepreneurialism: the transformation in urban governance in late capitalism. *Geografiska Annaler. Series B, Human Geography*, **71**(1), 3–17.

He, S., Li, Z, and Wu, F. (2006) Transformation of the Chinese city, 1995–2005: geographical perspectives and geographers' contributions. *China Information*, **20**(3), 429–456.

Heath, T. (2001) Adaptive re-use of offices for residential use: the experiences of London and Toronto. *Cities*, **18**(3), 173–184.

Helms, A.C. (2003) Understanding gentrification: an empirical analysis of the determinants of urban housing renovation. *Journal of Urban Economics*, **54**(3), 474–498.

Henderson, J.V. (2007) Understanding knowledge spillovers. *Regional Science and Urban Economics*, **37**(4), 497–508.

Hillier, B. and Hanson, J. (1984) *The Social Logic of Space.* Cambridge University Press: Cambridge.

Hutton, T.A. (2004) The new economy of the inner city. *Cities*, **21**(2), 89–108.

Hutton, T.A. (2006) Spatiality, built form, and creative industry development in the inner city. *Environment and Planning A*, **38**(10), 1819–1841.

Ignaccolo, M., Caprì, S., Giunta, U. and Inturri, G. (2006) Discrete choice model for defining a parking-fee policy on Island of Ortigia, Siracusa. *Journal of Urban Planning and Development*, **132**(3), 147–155.

International Energy Agency (2007) *Energy Use in the New Millennium: Trends in IEA Countries.* OECD/IEA: Paris.

Jacobs, J. (1961) *The Death and Life of Great American Cities.* Random House: New York.

Jencks, C. (2005) *The Iconic Building: The Power of Enigma.* Frances Lincoln: London.

Johnson, J.P. and Myatt, D.P. (2006) On the simple economics of advertising, marketing, and product design. *American Economic Review*, **96**(3), 756–784.

Kalay, Y.E. (2006) The impact of information technology on design methods, products and practices. *Design Studies*, **27**(3), 357–380.

Kellett, J.R. (1969) *The Impact of Railways on Victorian Cities.* Routledge & Kegan Paul: London.

Khalid, G. (1994) Obsolescence in hedonic price estimation of the financial impact of commercial office buildings: the case of Kuala Lumpur. *Construction Management and Economics*, **12**(1), 37–44.

Kohler, N. and Hassler, U. (2002) The building stock as a research object. *Building Research and Information*, **30**(4), 226–236.

Kohler, N. and Yang, W. (2007) Long-term management of building stocks. *Building Research and Information*, 35(4), 351–362.

Kok, H.J. (2007) Restructuring retail property markets in Central Europe: impacts on urban space. *Journal of Housing and the Built Environment*, 22(1), 107–126.

Kostof, S. (1992) *The City Assembled: The Elements of Urban Form Through History*. Thames & Hudson: London.

Kotus, J. (2006) Changes in the spatial structure of a large Polish city – The case of Poznań. *Cities*, 23(5), pp. 364–381.

Kristensen, T. (2004) The physical context of creativity. *Creativity and Innovation Management*, 13(2), 89–96.

Leaman, A. (2003) User needs and expectations. In: Cole, R.J. and Lorch, R. (eds) *Buildings, Culture and Environment: Informing Local and Global Practices*, pp. 154–176. Blackwell: Oxford.

Leamer, E.E., and Storper, M. (2001) The economic geography of the Internet Age. *Journal of International Business Studies*, 32(4), 641–665.

Liang, J., Li, B., Wu, Y. and Yao, R. (2007) An investigation of the existing situation and trends in building energy efficiency management in China. *Energy and Buildings*, 39(10), 1098–1106.

Lucas, R.E., Jr. and Rossi-Hansberg, E. (2002) On the internal structure of cities. *Econometrica*, 70(4), 1445–1476.

Lützkendorf, T. and Lorenz, D. (2007) Integrating sustainability into property risk assessments for market transformation. *Building Research and Information*, 35(6), 644–661.

Lützkendorf, T. and Speer, T. (2005) Alleviating asymmetric information in property markets: building performance and product quality as signals for consumers. *Building Research and Information*, 33(2), 182–195.

Markus, T.A. (1993) *Buildings & Power: Freedom and Control in the Origin of Modern Building Types*. Routledge: London.

Marshall, A. (1920) *Principles of Economics*. Macmillan: London.

McQuire, S. (2003) From glass architecture to Big Brother: scenes from a cultural history of transparency. *Cultural Studies Review*, 9(1), 103–123.

Meikle, J.L. and Connaughton, J.N. (1994) How long should housing last? Some implications of the age and probable life of housing in England. *Construction Management and Economics*, 12(4), 315–321.

Méndez, F. (2006) The value of legal housing titles: an empirical study. *Journal of Housing Economics*, 15(2), 143–155.

Moss, M.L. and Townsend, A.M. (2000) How telecommunications systems are transforming urban spaces. In: Wheeler, J.O., Aoyama, Y. and Warf, B. (eds) *Cities in the Telecommunications Age: The Fracturing of Geographies*, pp. 31–41. Routledge: New York.

Muhammad, S., Ottens, H.F.L., Ettema, D. and de Jong, T. (2007) Telecommuting and residential location preferences: a case study of the Netherlands. *Journal of Housing and the Built Environment*, 22(4), 339–358.

Müller, D.B. (2006) Stock dynamics for forecasting material flows: case study for housing in the Netherlands. *Ecological Economics*, 59(1), 142–156.

Mumford, L. (1961) *The City in History: Its Origins, Its Transformations, and Its Prospects*. Harcourt, Brace Jovanovich: New York.

Munoz, F. (2003) Lock living: urban sprawl in Mediterranean cities. *Cities*, 20(6), 381–385.

Nässén, J. and Holmberg, J. (2005) Energy efficiency – a forgotten goal in the Swedish building sector. *Energy Policy*, 33(8), 1037–1051.

Nelson, P. (1974) Advertising as information. *Journal of Political Economy*, 82(4), 729–754.

Nuissl, H. and Rink, D. (2005) The 'production' of urban sprawl in eastern Germany as a phenomenon of post-socialist transformation. *Cities*, 22(2), 123–134.

O'Flaherty, B. (1993) Abandoned buildings: a stochastic analysis. *Journal of Urban Economics*, 34(1), 43–74.

Phang, S.-Y. and Toh, R.S. (2004) Road congestion pricing in Singapore: 1975 to 2003. *Transportation Journal*, 43(2), 16–25.

Pickerill, T. and Pickard, R. (2007) A review of fiscal measures to benefit heritage conservation. *RICS Research paper series*, 7(6).

Rappaport, J. and Sachs, J.D. (2003) The United States as a coastal nation. *Journal of Economic Growth*, 8(1), 5–46.

Riley, J.G. (2001) Silver signals: twenty-five years of screening and signaling. *Journal of Economic Literature*, 39(2), 432–478.

Rossi-Hansberg, E. (2004) Cities under stress. *Journal of Monetary Economics*, 51(5), 903–927.

Sawyer, S., Wigand, R.T. and Crowston, K. (2005) Redefining access: uses and roles of information and communication technologies in the US residential real estate industry from 1995 to 2005. *Journal of Information Technology*, 20(4), 213–223.

Schiller, G. (2007) Urban infrastructure: challenges for resource efficiency in the building stock. *Building Research and Information*, 35(4), 399–411.

Searle, G. and De Valence, G. (2005) The urban emergence of a new information industry: Sydney's multimedia firms. *Geographical Research*, 43(2), 238–253.

Stamp, G. (2007) *Britain's Lost Cities*. Aurum: London.

Stern, N. (2006) *The Economics of Climate Change: The Stern Review*. Cambridge University Press: Cambridge.

Tether, B.S. and Metcalfe, J.S. (2003) Horndal at Heathrow? Capacity creation through co-operation and system evolution. *Industrial and Corporate Change*, 12(3), 437–476.

Thompson, B. (2005) Information and communications technology and industrial property. *Journal of Property Investment & Finance*, 23(6), 506–515.

Thurley, S. (2007) Wake up: Britain is being demolished under our very noses. *The Spectator*, 17 November, 14–15.

Toker, U. and Gray, D.O. (2008) Innovation spaces: workspace planning and innovation in U.S. university research centres. *Research Policy*, 37(2), 309–329.

Turok, I. and Mykhnenko, V. (2007) The trajectories of European cities, 1960–2005. *Cities*, 24(3), 165–182.

UN-Habitat (2003) *The Challenge of Slums: Global Report on Urban Settlements 2003*. Earthscan: London.

UN-Habitat (2006) *The State of the World's Cities Report 2006/2007*. Earthscan: London.

Van der Lugt, R., Janssen, S., Kuperus, S. and de Lange, E. (2007) Future center 'The Shipyard': learning from planning, developing, using and refining a creative facility. *Creativity and Innovation Management*, 16(1), 66–79.

Vesselinov, E., Cazessus, M. and Falk, W. (2007) Gated communities and spatial inequality. *Journal of Urban Affairs*, 29(2), 109–127.

Weber, R. (2002) Extracting value from the city: neoliberalism and urban redevelopment. *Antipode*, **34**(3), 519–540.

Wu, F. (2004) Transplanting cityscapes: the use of imagined globalization in housing commodification in Beijing. *Area*, **36**(3), 227–234.

Wu, J. (2006) Environmental amenities, urban sprawl, and community characteristics. *Journal of Environmental Economics and Management*, **52**(2), 527–547.

Xu, J. and Yeh, A.G.O. (2005) City repositioning and competitiveness building in regional development: new development strategies in Guangzhou, China. *International Journal of Urban and Regional Economics*, **29**(2), 283–308.

Xue, C.Q.L. and Zhou, M. (2007) Importation and adaptation: building 'one city and nine towns' in Shanghai: a case study of Vittorio Gregotti's plan of Pujiang Town. *Urban Design International*, **12**(1), 21–40.

Yin, M. and Sun, J. (2007) The impacts of state growth management programs on urban sprawl in the 1990s. *Journal of Urban Affairs*, **29**(2), 149–179.

Yonemoto, K. (2007) Endogenous determination of historical amenities and the residential location choice. *Annals of Regional Science*, **41**(4), 967–993.

Chapter 2

Quantifying the GDP–construction relationship

Timothy Michael Lewis

Introduction

Even to the casual observer it would seem likely that there is some sort of fairly direct relationship between the level of activity in the construction sector and in the economy as a whole. When the construction sector is booming, so too is the economy; when the economy is down, so too is construction. But it is not clear whether this is a coincidence or a causal relationship. If it is a causal relationship, which is the driver and which the follower? This is clearly important in policy terms if the objective is to achieve economic growth and development. In the same way, it is clear that the developmental status of the country has some sort of relationship with the role and size of the construction industry.

It has been said many times that construction is important in both financial and employment terms. Before looking at the industry statistics, it is important to note that the *construction industry* is normally narrowly defined to involve only those firms directly involved in the erection of structures and other constructed facilities. However, the industry can also be seen to be part of a wider *construction sector* that involves all aspects of the business from obtaining the raw materials through to demolition and disposal of the facility at the end of its useful life. The figures included in national economic statistics are generally based on the narrowly defined construction industry, however, various countries choose to include some or all of these 'support' elements in their definition of the construction industry, and as a result the statistics are rarely directly comparable. Because of this, historically, construction statistics have generally been poor and inconsistent, and this has made detailed studies of construction difficult – as Ruddock and Lopes (2006) note: 'A major obstacle to such studies has been the lack of appropriate information . . . particularly in developing countries.'

There is a movement towards standardizing the approach to such issues and one such mover is the Economic Statistics and Classifications Section of the United Nations. This organization indicates that statistics on construction include 'general construction and specialized construction activities for buildings and civil engineering works. It includes new work, repair, additions and alterations, the erection of prefabricated buildings or structures on the site and also construction of a temporary nature. General construction is the construction of entire dwellings, office buildings, stores and other public and utility buildings, farm buildings etc., or the construction of civil engineering works such as motorways, streets, bridges, tunnels, railways, airfields, harbours and other water projects, irrigation systems, sewerage systems, industrial facilities, pipelines and electric lines, sports facilities etc. This work can be carried out on 'own account' or on a fee or contract basis. Portions of the work and sometimes even the whole practical work can be subcontracted out. A unit that carries the overall responsibility for a construction project is classified here. Also included is the repair of buildings and engineering works. This section also includes the development of building projects for buildings or civil engineering works by bringing together financial, technical and physical means to realize the construction projects for later sale.'

That seems fairly comprehensive, until it is realized that there are quite a few activities that might have been expected to be included. For example, construction does **not** include:

- Architectural and engineering activities and related technical consultancy including:
    - Project management activities related to construction.
    - Building design and drafting.
    - Town and city planning and landscape architecture.
    - Engineering and consulting activities for projects involving civil engineering, hydraulic engineering, traffic engineering or water management projects.
    - Geophysical, geologic and seismic surveying.
    - Geodetic surveying activities.
- Mining and quarrying of materials used in construction.
- Manufacture of wooden goods for use in the construction industry.
- The manufacture of articles of concrete, cement and plaster for use in construction.
- The cutting, shaping and finishing of stone for use in construction.
- Manufacture of structural metal products (such as metal frameworks or parts for construction).
- Manufacture of machinery for mining, quarrying and construction.
- Construction of floating structures.
- Construction of drilling platforms, floating or submersible.
- Repair and maintenance of construction equipment and machinery.

- Collection and disposal of construction and demolition waste.
- Renting and operational leasing of construction and civil engineering machinery and equipment without operator (but 'construction' includes these when they are rented with the operator).
- Services to buildings and landscape activities such as the interior and exterior cleaning of buildings.
- Disinfecting and exterminating activities for buildings.
- Provision of landscape care and maintenance services and provision of these services along with the design of landscape plans and/or the construction (i.e. installation) of walkways, retaining walls, decks, fences, ponds, and similar structures.

Thus, despite the fact that construction uses huge quantities of certain materials and components which are manufactured for use exclusively by the construction industry, national accounts normally include firms involved in the manufacturing sector, not in construction (Kirmani, 1988). As a guide, the broadly defined construction *sector*, which includes all these elements, is estimated to be of the order of double the size of the construction *industry* as it is normally measured in the statistics (Pearce, 2003).[1]

In addition to the problems with what is included and what excluded from the figures, there are other reasons why construction has tended to be neglected by economists that are more due to the characteristics of the industry itself:

1   Construction is carried out by a large number of small contractors many of whom are not registered and who do not enter national or regional statistics, particularly in the developing countries.
2   A large proportion of the construction work is carried out by the informal sector and again is not included in national or regional statistics, and again particularly in the developing countries.
3   Construction is involved with all sectors of an economy and so it is not seen as a separate sector – because its outputs are investments or capital goods they are wanted, not so much for their own sake as for the goods and services they create, and so the focus is on the facility, not on its construction.
4   The practice of sub-contracting creates the problem of possible double counting.
5   Value added by construction is defined as the difference between the market value of the goods it produces and the value of all the goods and services it purchased from other sources. This is difficult to determine because of wage structures and patterns, plant shared between sites, the ways in which overheads and profits are assigned, and because accounts of the values of various inputs and outputs are not adequately prepared, particularly in developing countries.
6   Problems separating out the construction from public sector investment

plans and the difficulty of fully accounting for the value of maintenance and repairs (which can amount to as much as one third of the total output of construction).
7   The World Bank and other similar agencies have 'also not considered construction an important development issue' (Kirmani, 1988, p. 31).

Before we go on to look at the detals of the scale of construction in relation to the economy, we will look at a little background on studies of this relationship.

The role of the construction industry

In the late 1960s and early 1970s, Duccio Turin (1969) and Paul Strassmann (1970) were independently, focusing on the role and status of the construction[2] industry in national economic development. Both were interested in examining the relationship between the level of output of the economy as a whole and activity in construction, as well as the relationship between the nature of the construction industry and the economic development of the country concerned. Specifically Strassman (1970, p. 391) was interested in the answer to the question 'Does the construction sector, like agriculture or manufacturing, follow a pattern of change that reflects a country's level of development?' An answer to this question can be found by looking at the statistics of the industry and the economy as a whole for a range of countries at different stages of development. For ease of analysis income per capita or gross domestic product per capita are used as proxy measures of 'development' – with higher levels obviously indicating a more developed status.

Turin (1969) and Strassman (1970) were both interested in the place of construction in the national economy and approached the study by relating 'the main aggregates of construction activity to the level of economic development'. The level of economic development is largely determined by a set of economic characteristics such as the gross national (or domestic) product (GNP or GDP), population and employment figures, rate of investment, and other indices of economic activity that are published by many countries. As Leontief (1965) wrote 'Such figures give quantitative expression to the otherwise plainly apparent fact that some countries are rich and others poor. When these figures have been plotted over the recent past, they indicate that the gap between the rich and the poor has been widening.'

Economists since Rostow[3] (1956, 1988) have tended to believe that economic growth is a process with a number of stages that have to be followed, although their duration may be different. Thus, even though the current levels of development may be different, all countries are working their way along the same development path, and as a result there are lessons that can be learnt from the developed countries by those that are developing. One of the patterns of development relates to the construction industry and its relationship with the economy as a whole.

One of the tools used to analyse the relationships between industries in an economy is the input–output table, which measures the relationships in terms of who bought and sold what, and from whom. The performance of an economy is 'determined by the mutual relations of its differentiated component parts, just as the motion of the hands of a clock is governed by the gears inside'. If you want to fix the clock you need to know how the parts fit together, and to understand what happens if you change the size of one gear in relation to the others. The cells in an input–output table record the transactions between industries and an empty cell indicates that that particular buying industry made no purchases from that particular selling industry. The more empty cells, the less developed the industries in the economy and the more dependent that the country is on inputs from abroad. For the developing countries, their 'input–output tables show that in addition to being smaller and poorer they have internal structures that are different because they are incomplete, compared with the developed economies' (Leontief, 1965). Thus, the larger and the more developed an economy is the more 'complete and articulated is its structure'. Each cell with an entry identifies a linkage between the two industries or sectors involved, and these linkages are defined according to the direction of the interdependency. A backward linkage is one that identifies where one industry purchases its inputs, and a forward linkage identifies where the industry sells its outputs.

Obviously, a country can use or consume inputs without producing them because it can import them, but this is at a cost to the economy. The smaller and less developed a country is, the more it is likely to be dependent on foreign trade for a wide range of goods and factor inputs.

Turin assumed that it is reasonable to summarize the main economic characteristics of construction output over a group of countries at the same level of social and economic development and to take this as being representative for all countries at a similar level. Thus, the characteristics of the average values for all developing countries can be accepted as typical for any developing country, and similarly for developed countries.

Turin (1969) noted that there are many inconsistencies in the national accounts statistics that are used to compare economic performance, particularly from the developing countries,[4] but that aggregating the data helped remove some of the reservations and allowed patterns to be identified. This helped highlight the role of construction in a national economy, and a number of commentators[5] have built on this work since then. Indeed, Strassman (1970) felt that construction had overtaken manufacturing as the driving force for economic growth in countries that had begun the process of economic development. A number of studies[6] have attempted to use input–output analysis to establish more explicitly the relationships of construction with other sectors; they have suffered from the inadequacy of the data available but in general have found strong backward and forward linkages with many other sectors. Strong linkages imply that whatever happens to construction will directly influence other industries and, ultimately, the economic wellbeing of the country.

Bon (2000) reports on studies of 'highly developed economies', including the USA, Japan, Italy, UK, Ireland and Finland. These studies show that the construction industry in these countries is characterized by the trend[7] for the industry's economic importance to follow an 'inverted U' pattern of growth followed by decline with increasing national economic development. In addition to its relationship to the developmental status of a country, construction also appears to have a fairly direct relationship with the national output (Field and Ofori, 1988) – when construction is down, the economy, almost inevitably, will also be down; when it is up, so too will be the economy.

Turin's (1973) data suggested that there was a direct relationship between the level of construction activity and per capita GDP, and that, in the developing countries, value added by construction was growing faster than per capita GDP.[8] It was also apparent that there were significant differences in the size of the construction sector in relation to the economy in developed and developing countries. In most developing countries, construction accounted for between 3 and 5 per cent of GDP whilst in developed, industrialized countries it accounted for almost twice as much (between 5 and 9 per cent) (Ruddock, 1999). Recent figures suggest that this is no longer the case, as will be shown below.

Turin's figures in general show that the share of construction in the national product and the value added in construction per capita grow with economic development. More recent studies have suggested that construction's share of GDP declines after a certain level of economic development – the 'inverted U' model (Bon, 1992; Crosthwaite, 2000).

Tan (2002) explains this phenomenon: 'In low income countries (L), construction output is low. As industrialization proceeds, factories, offices, infrastructure and houses are required, and construction output as a percentage of gross domestic product (GDP) reaches a peak in middle income countries (M). It then tapers off in high income countries (H) as the infrastructure becomes more developed and housing shortages are less severe or are eliminated.' In the words of Bon and Pietroforte (1990) 'the decline of the construction sector sets in with economic maturity: the more developed an economy, the smaller the construction sector' (see also Pietroforte and Gregori, 2006).

Because the government plays an important role as a purchaser of construction, it can significantly affect prices, resource allocation and output decisions. This is particularly the case in the area of infrastructure especially in countries at the earlier stages of development and, as Turin (1969) highlights, the share of expenditure in the developing countries directed toward construction of infrastructure is typically greater than in more developed economies (see also Birgonul and Ozdogan, 1999).

Empirical work on the relationship between the construction sector and the level of economic activity in the rest of the economy, has generally been based on Keynesian economics via statistical correlates between a measure of construction output and one for national output. These studies identified that the share of construction in overall GDP takes the shape of an inverted U as the developmental status of the country matures.

By way of illustration (Figure 2.1) of this relationship, if Y_i (i = l, m, h) represents low, medium and high income respectively, then generally at Y_1, $(^c/_y)_L$ prevails. As we move from Y_1 to Y_m the $^c/_y$ ratio expands to $(^c/_y)_M$. Beyond Y_m, however, the $^c/_y$ ratio decreases (note $(^c/_y)_H$ can fall below $(^c/_y)_L$ although practically with the high costs of repair and maintenance would not – the main point is that overall $(^c/_y)_H < (^c/_y)_M$). Empirical support for this type of work was presented in research work by Strassman (1970) and Turin (1973).[9]

Construction's role contracts in the economy after the Y_m stage as the industry's relative wages expand and its productivity falls. Turin put forward his argument on the basis of Rostow's stages of economic growth argument. In particular, Turin argued that countries in the earlier part of their developmental ladder attribute a larger element of their resources to new work particularly the development of the basic infrastructure, in agriculture, mining and transportation and communication. These economies also need to expend a considerable amount of resources on social sector formation such as education and health-care facilities.

The lack of proper maintenance of work in any country would mean that the physical assets would deteriorate, and in order to avoid this sufficient funds have to be budgeted for repairs, renovations, refurbishment and other general maintenance. Present day decisions on the quality and the quantity of new work and on the level of maintenance that will be carried out on the existing stock of structures implicitly commits future generations to a pattern of expenditure, which will almost inevitably lean heavily on the side of maintenance and repairs (Turin, 1973, p. 4).

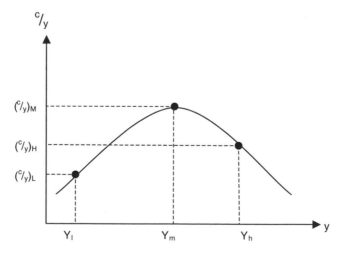

Figure 2.1 The inverted U diagram.

Figures published recently[10] allow us to examine the role of construction in a national economy, starting with its contribution to a nation's Gross Domestic Product (GDP). As illustrated in Figure 2.2, the figures show that construction has been responsible for around 6 per cent of GDP of a selection of developed countries over the 36 years examined. It will also be noted that in the USA, for example, the constant dollar[11] value of the share of construction in GDP has drifted down from around 7 per cent to around 4 per cent between 1970 and 2006. In Japan the decline was much more dramatic, falling from over 14 per cent to less than 6 per cent. The same trend is followed by all of the developed countries examined.

These figures suggest that the industry is relatively small and declining in importance nowadays. This characteristic was predicted by a number of earlier studies (Wells, 1986; Bon, 1992) in the form of the inverted U – the prediction that share of construction in GDP would grow as a country begins development, but then, after a certain level is reached, begin to decline again. As a result, on the basis that these countries are already developed, it would be expected that the share of construction in their GDP would be declining.

It should, however, also be noted that these statistics (for the construction *industry*) will significantly under-represent the true scale of the construction *sector*. However, while the scale of its contribution may be less than it should be, it is unlikely that the trend would be different.

The outriders in a figure like this are usually the ones that are most instructive as they may not only illustrate the pattern but also exaggerate it. The predictions based on the 'inverted U' model suggest that there would be a fall off in the demand for construction as a country becomes more developed – and should already have provided most of the infrastructure and buildings

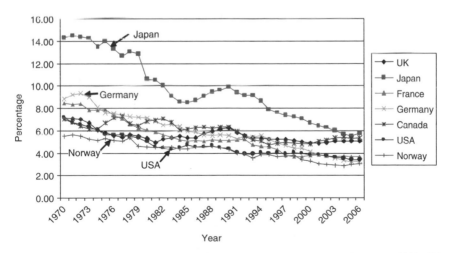

Figure 2.2 Construction as percentage of GDP in developed countries (constant 1990 US$ values).

that it requires. Hence, the trend shown in Figure 2.2 is understandable. However, the drop in the share of construction in national GDP shown by Japan appears larger than might have been expected. This requires some explanation, and it comes from Figure 2.3, where it can be seen that in Japan construction has stayed fairly steady in terms of dollar value, whilst the economy has grown significantly. Hence, the fall off in construction's share of GDP reflects a more than proportional growth in GDP rather than an absolute decline in construction. Detailed analysis reveals that most of the developed countries follow the same trend, with national GDP growing faster than construction output. The USA is slightly different where, as Figure 2.4 shows, although there is a steady growth trend, the pattern is rather unsettled for the economy as a whole as well as for construction. The two combine to produce a fairly stable contribution of construction to the economy as a whole.

Figure 2.5 shows the share of the construction industry in the economies of a selection of developing countries over the same time period. Although the overall trend is basically similar, the pattern amongst the developing countries is not so consistent as for the developed countries. A number of the countries show the share of construction in GDP varying around a mean and Guyana actually shows a slow but steady growth. Some of the countries show fairly dramatic swings from growth to decline, with Trinidad & Tobago being a particular case in point. Given the predictions of the 'inverted U' model, it could be suggested that Guyana is still below the 'turning point' in development at which construction starts to decline, and that countries like Trinidad & Tobago are passing through that stage now.

Figure 2.5 shows the share of construction in GDP, but to examine the

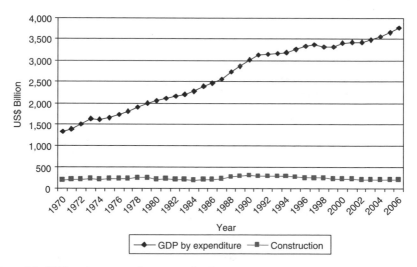

Figure 2.3 GDP and construction output for Japan.

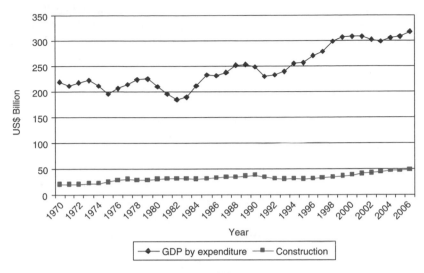

Figure 2.4 **GDP and construction output for USA.**

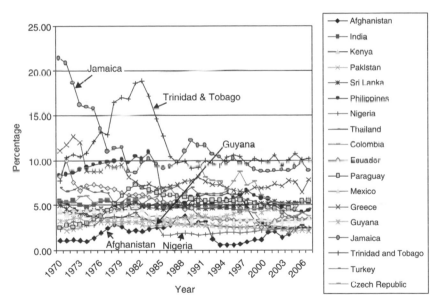

Figure 2.5 **Construction as percentage of GDP in developing countries (constant 1990 US$ values).**

implications of these shapes it is necessary to go back to the actual values of construction output and GDP to see what they show. What the curves actually show is that, in general, the output of the construction industry has been fairly steady in real terms, but that the national economies have been growing.

Again it is instructive to look at the outliers, and those, if any, that 'buck the trend'. The low outliers are Afghanistan and Nigeria. Figure 2.6 shows that Afghanistan's economy has been very unsettled, and construction has continued at a low level. The scale of the contribution of construction is affected more by the variations in the national economy than by changes in the output of the industry. The same applies to Nigeria, another of the low outliers, as shown in Figure 2.7.

Figures 2.8 and 2.9 show the curves for Jamaica and Trinidad & Tobago, which are the high level outliers in the grouping. The share of construction in GDP for Jamaica shows a significant decline between the start and end of the period with some fluctuations in between. This may suggest that it has reached the stage of economic development at which the share of construction in the economy starts to decline, but familiarity with the country would suggest otherwise. Jamaica did reach a point of reasonable development, but still significantly lacks the sort of infrastructure and building stock for industrial and

Figure 2.6 Afghanistan.

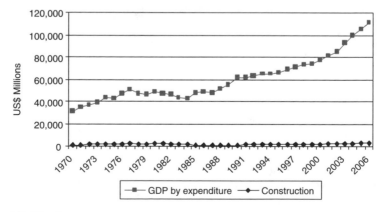

Figure 2.7 Nigeria.

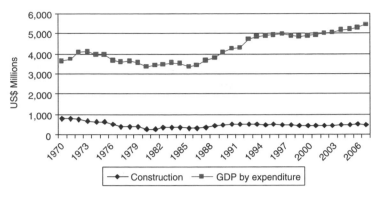

Figure 2.8 Jamaica.

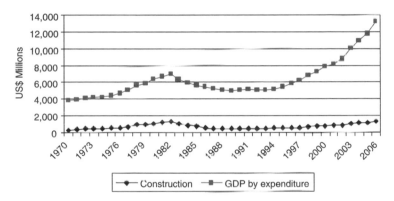

Figure 2.9 Trinidad & Tobago.

economic development. The decline in the share of construction in GDP reflects something other than developmental maturity.

As Figure 2.10 shows, Jamaica has not shown any growth in value added by the construction industry or per capita in the industry in real terms over the 30 odd years under consideration. Indeed the value added per capita still shows no signs of significant improvement. This is at least in part explained by the population levels have kept growing, having increased from around 1.9 million to around 2.7 million over the period. Thus, any growth in GDP has been offset by the increaase in population – to the extent that there has been little opportunity for improvement in standards of living.

The other outlier on the high side is Trinidad & Tobago.

One country that has been out of step with most of the others is Guyana (Figure 2.11), where the growth of the construction industry has recently outstripped that of the economy, so that the share of construction in GDP has grown from under 4 per cent to over 5 per cent of GDP. This is partly explained

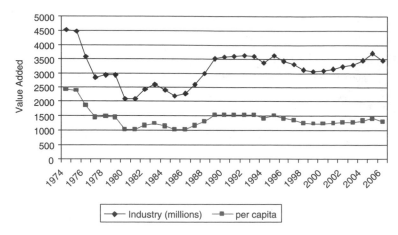

Figure 2.10 Jamaica construction value added (constant 1990 US$).

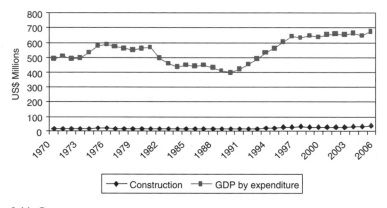

Figure 2.11 Guyana.

by the sluggish growth of the economy as a whole but also by the absolute growth of construction as the country determinedly builds its infrastructure, commercial and social facilities in its search for improved developmental status.

A number of developing countries have recently favoured providing infrastructure through involvement of private capital, usually in the form of concessions which involve the private developers in some form of build, own and operate contract. Where such facilities are designed and constructed by foreign firms they would not show up in the construction industry data, and depending on how they were financed, they may not show up in GDP figures, and these are potentially large transactions that would significantly impact the statistics.

The rate at which a country can develop is restricted by its ability to use its domestic resources effectively. This is determined by the quality and scale of its investment programme. Because construction is the main industry charged with implementing the investment programme, the capacity and efficiency of the industry are critical to development. Increasing industrialization tends to require higher levels of investment, and so increasing levels of investment can indicate increasing development. The investment that underpins industrialization and economic development is almost entirely 'capital formation' created by the construction industry,[12] and so '. . . it is therefore to be expected that whenever such growth occurs it must be accompanied by a rapid expansion of activity in the construction sector. Thus common sense tells us there must be a close relationship between construction output and economic growth.' (Wells, 1985) When national income is high and increasing, a large part of the income is devoted to investment in fixed assets (e.g. dams, roads, factories, schools, housing, etc.), all of which involve construction. When national income is reduced, investment also decreases. As higher income countries can save more than lower income countries they are able to invest more. 'Thus, the higher the national income, the greater the size of investment, the volume of construction activity, the increase in GDFC (productive capital stock) and the growth in GDP. This chain of relationships underlines the important place of the construction sector in economic development' (Kirmani, 1988).

Historically, construction accounted for around 50 per cent of a country's gross fixed capital formation (GFCF).[13] Figures from the UN statistical service suggest that this is rather high and that nowadays, for the developed countries the figure is nearer 26 per cent and for the developing countries it is around 23 per cent. Figure 2.12 illustrates the changes over time of the proportion of construction in GFCF for the developed countries over a period of some 36 years. It will be seen that the lines appear to be converging at a little below 20 per cent with a range of 10 percentage points from 12.4 to 22.4 per cent.

The outliers in Figure 2.12 are Japan and Norway. Although it tends to be

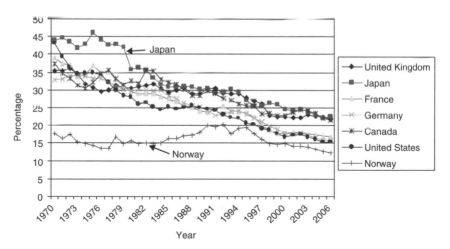

Figure 2.12 Construction as percenatge of GFCF in developed countries.

higher than any of the other curves for almost the whole of this period, Japan's trend is characteristic of the group. Norway on the other hand is quite different, and Figure 2.13 illustrates the curve for Norway. Although the overall trend is towards a decreasing share of construction in GFCF, it shows a very variable relationship.

Figure 2.14 shows the same relationship for the developing countries.

Figures 2.15 and 2.16 show the share of construction in GFCF for Trinidad & Tobago and Nigeria respectively. It will be apparent that the general shapes of the curves are very similar, though their range is different, with the former climbing above 200 per cent while the latter barely makes it above 50 per cent.

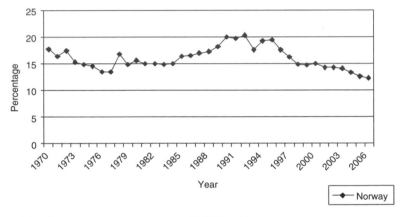

Figure 2.13 Construction as percentage of GFCF in Norway.

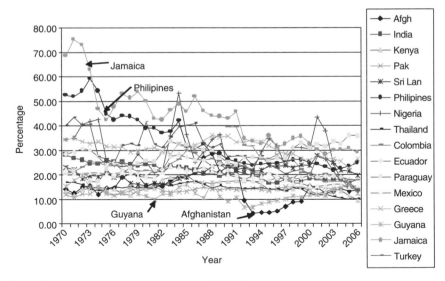

Figure 2.14 Construction as percentage of GFCF in developing countries.

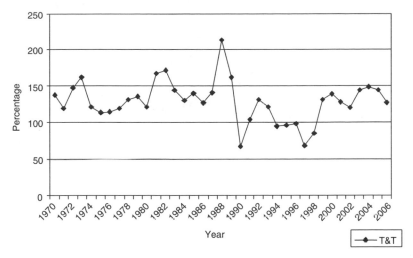

Figure 2.15 Construction as percentage of GFCF in Trinidad & Tobago.

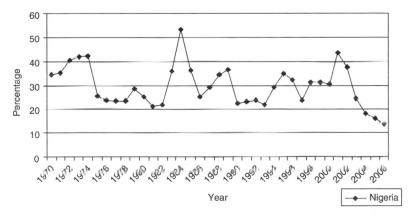

Figure 2.16 Construction as percentage of GFCF in Nigeria.

Low (1991b) found that, typically, in developing countries capital formation in construction accounts for a lower proportion of GDP than in developed countries (7–13 per cent compared with 10–16 per cent). This does not seem to be borne out by the figures reported here. Although at the bottom end of the scale, construction does account for a smaller share of GFCF than in the developed countries, at the top end, it also accounts for a higher share.

Because of its role as a provider of the capital assets, construction's share of GFCF is an important measure of its contribution to economic development, obviously this is conditioned by the scale of the GFCF in relation to the size of the economy. Figure 2.17 shows the relationship between GFCF and GDP for the same set of developed countries. As will be seen, the trend has been

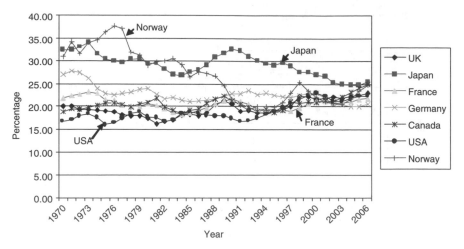

Figure 2.17 GFCF as percentage of GDP in developed countries.

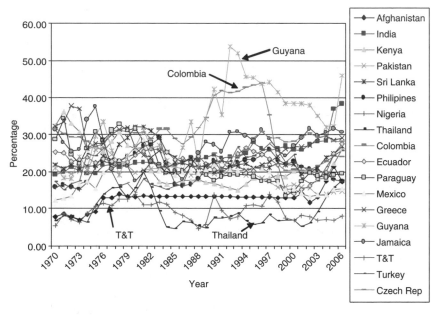

Figure 2.18 GFCF as percentage of GDP in developing countries.

for countries to converge on a value around 23 per cent, within a range of 22–26 per cent. In other words, the creation of new fixed capital represents around one quarter of the GDP of these countries.

Figure 2.18 shows a somewhat different picture for the developing countries where there appears a tendency to diverge, so that the share of fixed capital formation in GDP has tended to become less and less similar across countries over time. One obvious reason for this is that government expenditure is normally responsible for much of the fixed capital in the form of infrastructure works and public buildings. These expenditures are the first to be cut back in times of economic difficulty, such as during the international recession of the early 1980s. As can be seen from the diagram, in 2006 GFCF represented between 8 and 46 per cent of GDP, with an average of around 20 per cent.

The somewhat erratic relationship between the share of GFCF and GDP for Thailand does not show up in the chart of GFCF against GDP in dollar terms. Figure 2.19 shows a plot of GDP and of GFCF for the period 1970–2006, and the two curves seem to follow one another quite closely, especially in registering the economic downturn between 1997 and 2000. Since around 2003 the share of GFCF in GDP has indcreased significantly, climbing from just over 5 per cent to just over 17.5 per cent.

The low level erratic relationship shown by Thailand is also characteristic of Trinidad & Tobago (as shown in Figure 2.20), except that the share of GFCF in GDP starts at just under 7 per cent and finishes at just over 8 per cent. When the absolute values are plotted as shown in Figure 2.20 it can be seen that although there has been significant growth in GFCF, the GDP has been growing even faster, and this has depressed the level of contribution GFCF has made in recent times. It has to be said that the GFCF seems low considering the heavy investments being made in oil and gas processing plants, and infrastructure and commercial, public and residential buildings at the present time. It may be that an investment by a foreign company in a gas processing facility that is being built predominantly by a foreign contractor would not show up in the GFCF figures for Trinidad & Tobago, as both declare their earnings outside the country.

At the other end of the scale, there are two high outriders that are of interest. For much of the 1990s GFCF accounted for around 43 per cent of GPD for Colombia. It had climbed there from a low of around 27 per cent in the

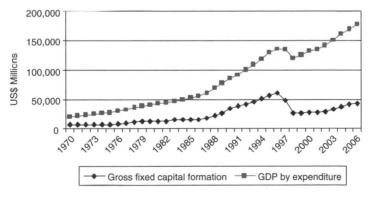

Figure 2.19 Thailand.

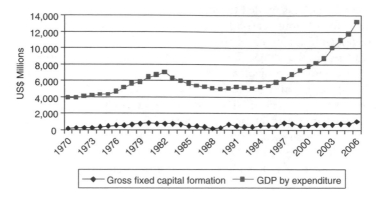

Figure 2.20 Trinidad & Tobago.

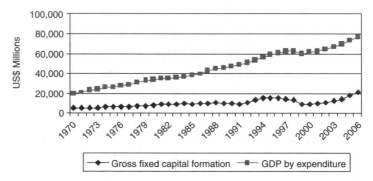

Figure 2.21 Colombia.

mid-1970s, and it rapidly fell back to below these levels within a couple of years between 1996 and 1998. It has recovered slightly to around 24 per cent of GDP currently. These rapid changes in GFCF do not show up in the figures for the dollar values of GFCF and GDP. As is shown in Figure 2.21 GFCF and GDP seem to progress more or less in step.

The other high outrider is Guyana. In the early 1990s, Guyana had a share of GFCF in GDP approaching 54 per cent. That is to say almost half of the country's GDP was capital formation. Despite a retreat to around 32 per cent during the mid 1990s and early 2000s, it has climbed again to around 46 per cent of GDP. This is still a very significant share of GPD that is going to GFCF.

The curve in Figure 2.22 shows that economic progress and GFCF have both had a bumpy ride over the period shown. Guyana has an economy that is dependent on commodity prices over which it has no control, and which have proved very volatile over this period. It is not likely that this will change much in the near future, though it is possible that the economy will benefit from rising prices for its natural resources.

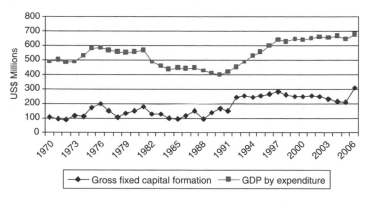

Figure 2.22 Guyana.

Causality

A recent study in Taiwan (Chang et al., 2004) examined the relationship between construction activity and economic growth and particularly the direction of causality. Their figures indicate that there is 'unidirectional causality running from construction activity to economic growth'. In other words, construction activity stimulates the economy rather than the other way round, so, in policy terms, if a country wants to boost its economy then one effective means is through construction activity.

Drewer (1997, p. 136) found that 'within a country through time, the relationship between construction and the wider economy is unstable and subject to significant variations'. He did, however, conclude that construction could effectively drive economic growth to a limited extent and for a relatively short period, but that the economy could be threatened by the spurt of economic growth resulting from a period of uncontrolled expansion of the construction industry. Such a period of construction industry expansion is partly blamed for the 1985 economic recession in Singapore (Economic Committee, 1986), as well as the economic problems of Trinidad & Tobago around the same time (Lewis, 1984).

Cumulative experience

In order to examine the direction of causality it is necessary to introduce the concept of the 'cumulative experience function'. The idea of the cumulative experience function provides non-parametric evidence on the direction of causality between two variables. Equation 1, below, outlines the format of a cumulative experience (cumexp) function for a variable x:

$$Cum \exp x = \sum_{i=t_0}^{t} x_t \Big/ \sum_{i=t_0}^{t_1} x_t \tag{1}$$

where t_0 and t_1 are the starting and finishing years of the data period and t is the current time period. The cumulative experience function adheres to distributional properties that are similar to the cumulative distribution function, i.e. $c_{it} \rightarrow$ 0 at values of $t = 0$ and towards 1 as $t \rightarrow t_1$, where t_1 is the terminal year of the data.

The data and the cumulative experience functions shown in Figure 2.23 were published by Lewis et al. (2004) and provide indicative evidence concerning the relationship between the level of economic activity, crude oil production and real oil revenues and activity in the construction sector of Trinidad & Tobago. This evidence suggests that real oil revenues and crude oil production led real construction value added for the time interval 1966–2002. The period 1966–1982 covers the period of booming economic activity associated with the OPEC raising of the price of oil and the consequent revenue increases to the Trinidad & Tobago national exchequer from oil exports. During this period, the value added of the construction sector was led by that of the overall macro economy. Between then and the mid-1990s construction led the economy. It is clear and logical that the relationship between construction and the national economy should change over time, and that under different circumstances one or the other will take the lead. The recent experience in Trinidad & Tobago has been that in the times of economic plenty, the economy leads construction particularly through increases in public sector demand for construction output, whilst in times of recession and economic hardship the construction industry drives the economy, and can be used to stimulate the economy when necessary.

Summary

The relationship between the ratio of the total output in construction to that in the economy as a whole shows that as an economy develops, construction

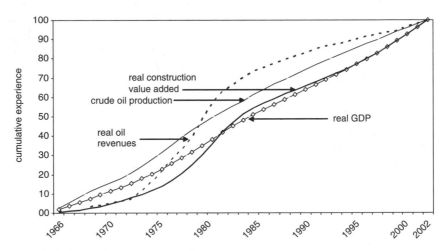

Figure 2.23 A cumulative experience function.

assumes a less important role in the economy. In the early stages of development construction can account for a large part of all economic activity, and can play a key role in modernization. This role is emphasized by the proportion of the construction spending that goes on 'infrastructure' – at lower levels of GNP per capita spending on infrastructure predominates – up to around 60 per cent of total capital formation; at higher levels of GNP per capita, this falls off to around 25–30 per cent. Economic modernization necessarily depends on the adequacy of the infrastructure and the general built environment.

It is possible that there may be some quite significant shifts in these ratios as many of the developed industrialized economies are facing the fact of a rapidly decaying installed infrastructure, and a massive bill for repair, renovation or replacement. Footing this bill, could change the balance of expenditures that have been seen over the past 50 years or so, with 'other construction and works' dominating over the 'dwellings' and 'non-residential buildings' – this may be especially the case with stable or possibly declining populations decreasing the demand for accommodation, with a large existing stock of houses already on the market.

Notes

1 That having been said, there are still a number of countries that follow a different system and include in the sector statistics not only the building, repair and maintenance of structures and facilities, but also '. . . a vast range of ancillary activities, such as the mining and manufacture of construction materials, the transport of such materials and equipment, the provision of professional services, like architecture, quantity surveying, and construction engineering' (Francis, 1997).

2 We will tend to focus here on the more narrowly defined construction industry, which involves 'Economic activity directed to the creation, renovation, repair or extension of fixed assets in the form of buildings, land improvements of an engineering nature and other such engineering constructions as roads, bridges, dams and so forth' (United Nations, 1997). The term 'construction sector' will be reserved for the more broadly defined discipline that includes, for example, all of the elements listed by Francis (1999).

3 Rostow suggested that a country must pass through five stages in achieving development: (1) the traditional stage with low and stagnant per capita output; (2) the transitional stage when the preconditions for growth are established; (3) the takeoff stage to economic growth; (4) the drive-to-maturity stage; and (5) the development stage of industrialized mass production and consumption.

4 Ruddock (1999) amongst others, such as Lopes (1998) and Francis (1997), bemoans the fact that, for the construction sector 'the availability of suitable, valid data has not improved over the past decade'.

5 These include Hillebrandt (1985), Wells (1985), Drewer (1980, 1990), Edmonds and Miles (1984), Field and Ofori (1988), Han and Ofori (2001) and Crosthwaite (2000).

6 A listing of texts on input–output analysis of the construction sector would include Lewis (1981), Bon and Minami (1986), Bon and Pietroforte (1990, 1993), Bon and Yashiro (1996), Bon (1988a,b, 1992, 2000), Pietroforte and Bon (1995), Pietroforte et al. (2000) Pietroforte and Gregori (2003, 2006), Lean (2001), Liu et al. (2005), Rameezdeen et al. (2005), Song et al. (2006).

7 Another trend identified by Bon was the progressive change in the industry's pro-duction system over time, in that its major inputs changed from manufactures and to service inputs. This is a similar pattern to the one followed by manufacturing itself some 20 years previously.

8 Because wages form a significant part of value added, this could have been because wages in construction were rising faster than the industrial norm.

9 It is very well established in the literature that correlation does not necessarily imply causation. Even more than this though if two variables X and Y are corre-lated, as well as causally linked, one would still need to determine if the causality is unidirectional from X to Y or Y to X or bidirectional i.e. from X to Y and Y to X.

10 The various analyses and figures produced and discussed here are derived from fig-ures published and available online from the United Nations Statistics Division, *National Accounts Main Aggregates Database*, http://unstats.un.org/unsd/snaama/status.asp (accessed 5 September 2007).

11 The constant dollar value is the current dollar value adjusted for the effects of inflation to its equivalent in 1990.

12 Sen (1983) for example notes that countries with high growth rates also showed high rates of capital accumulation.

13 The share of construction in GFCF does not appear to bear any direct relationship to the level of development, as measured by GDP *per capita* or any other variable; the only consistent pattern that has been observed is a falling share of construction in total investment in the centrally planned economies of Eastern Europe – perhaps a reflection of increased productivity in the construction sector (see Drewer, 1991), footnote from Wells (1985).

References

Birgonul, M.T. and Ozdogan, I. (1999) Government effects of the Turkish construction industry. In: Ruddock, L. (ed) *Macroeconomic Issues, Models and Methodologies for the Construction Sector*. CIB Publication No. 240: Rotterdam.

Bon, R. (1988a) Supply-side multiregional input–output models. *Journal of Regional Science*, 28, 41–50.

Bon, R. (1988b) Direct and indirect resource utilization by the construction sector: the case of the United States since World War II. *Habitat International*, 12, 49–74.

Bon, R. (1992) The future of international construction: secular patterns of growth and decline. *Habitat International*, 16(3), 119–128.

Bon, R. (2000) *Economic Structure and Maturity: Collected Papers in Input-Output Modelling and Applications*. Ashgate Publishing Ltd: Aldershot, England.

Bon, R. and Minami, K. (1986a) The role of construction in the national economy: a comparison of the fundamental structure of the US and Japanese input–output tables since World War II. *Habitat International*, 10(4), 93–99.

Bon, R. and Pietroforte, R. (1990) Historical comparison of construction sectors in the United States, Japan, Italy and Finland using input-output tables. *Construction Management and Economics*, 8(3), 233–247.

Bon, R. and Pietroforte, R. (1993) New construction versus maintenance and repair construction technology in the US since World War II. *Construction Management and Economics*, 11(2), 151–162.

Bon, R. and Yashiro, T. (1996) Some new evidence of old trends: Japanese construction 1960–90. *Construction Management and Economics*, 14(4), 319–323.

Cannon, J. (1994) Lies and construction statistics. *Construction Management and Economics*, **12**, 307–313.

Chang, T. and Nieh, C.-C. A note on testing the causal link between construction activity and economic growth in Taiwan. *Journal of Asian Economics*, **15**(3), 591–598.

Chen, J.J. (1998) The characteristics and current status of China's construction industry. *Construction Management and Economics*, **16**, 711–719.

Crosthwaite, D. (2000) The global construction market: a cross-sectional analysis. *Construction Management and Economics*, **18**(5), 619–627.

Dang dai zhong guo (1990) *Dang dai zhong guo de jing ji guan li (China Today: Economic Management)*. Dang dai zhong guo bian wei, Beijing (p. 245) quoted in Han and Ofori (2001).

Dikmen, I. and Birgonul, M.T. (2006) A review of international construction research: Ranko Bon's contribution. *Construction Management and Economics*, **24**(7), 725–733

Drewer, S. (1980) Construction and development: a new perspective. *Habitat International*, **5**(3/4), 395–428.

Drewer, S. (1990) The international construction system. *Habitat International*, **14**(2/3), 29–35.

Drewer, S. (1991) Technology for Development. In *MIMAR 38: Architecture in Development*. Concept Media Ltd: London.

Drewer, S. (1997) Construction and development: further reflections on the work of Duccio Turin. *Proceedings of the First International Conference on Construction Industry Development*, Singapore.

Edmonds, G. (1979) The construction industry in developing countries. *International Labor Review*, **118**(3), 355–369.

Edmonds, G. and Miles, D. (1984) *Foundations for Change: Aspects of the Construction Industry in Developing Countries*. ITP: London.

Economic Committee (1986) The Singapore economy: new directions. *Report of the Singapore Economic Committee*. Ministry of Trade and Industry: Singapore.

Field, B. and Ofori, G. (1988) Construction and economic development: a case study. *Third World Planning Review*, **10**(1), 41–50.

Francis, M. (1997) *Developments in the Construction Sector of Trinidad & Tobago: outlook for the 1990s*. Research report, pp. 63–72. Central Bank of Trinidad & Tobago: Port of Spain.

Han, S.S. and Ofori, G. (2001) Construction industry in China's regional economy, 1990–1998. *Construction Management and Economics*, **19**(2), 189–205.

Hillebrandt, P.M. (1985) *Economic Theory and the Construction Industry*, 2nd edition. Macmillan: Basingstoke.

Kirmani, S.S. (1988) *The Construction Industry in Development: Issues and Options*, Report INU 10, Infrastructure and Urban Development Department, p. 14. The World Bank: Washington.

Lean, S.C. (2001) Empirical tests to discern linkages between construction and other economic sectors in Singapore. *Construction Management and Economics*, **13**, 253–262.

Leontief, W. (1965) In: Briggs, A. (ed). *The Structure of Development, Technology and Economic Development*, pp.129–148. A Scientific American Book, Penguin: Harmondsworth.

Lewis, T.M. (1981) Input–output analysis of the construction industry in Trinidad &

Tobago. *Proceedings of 3rd International Symposium, CIB W-65, Organisation and Management of Construction*, Dublin. Volume 2, pp. 135–151.

Lewis, T.M. (1984) A review of the causes of recent problems in the construction industry of Trinidad & Tobago. *Construction Management & Economics*, 2, 37–48.

Lewis, T.M. and Hosein, R. (2004) Quantifying the relationship between aggregate GDP and construction value added in a small petroleum rich economy – a case study of Trinidad and Tobago. *Construction Management and Economics*, 22, 185–197.

Liu, C., Song, Y. and Langston, C. (2005) Economic indicator comparisons of multinational real estate sectors using the OECD input-output database. *The International Journal of Construction Management*, 4(1), 59–75.

Lopes, J. (1998) The construction industry and macroeconomy in Sub-Saharan Africa post-1970. *Construction Management and Economics*, 16(6), 637–649.

Low, S.-P. (1991a) World markets in construction 1: a regional analysis. *Construction Management and Economics*, 9(1), 63–71.

Low, S.-P. (1991b) World markets in construction 2: a country by country study. *Construction Management and Economics*, 9(1), 73–78.

Low, S.-P. (1994) Balancing construction and marketing in world economic development: the four global scenarios. *Construction Management and Economics*, 12(2), 171–182.

Pearce, D. (2003) *The Social and Economic Value of Construction: The Construction Industry's Contribution to Sustainable Development*, nCRISP. Davis Langdon Consultancy: London.

Pietroforte, R. and Bon, R. (1995) An input-output analysis of the Italian construction sector, 1959–1988, *Construction Management and Economics*, 13(3), 253–262.

Pietroforte, R. and Gregori, T. (2003). An input-output analysis of the construction sector in highly developed economies. *Construction Management and Economics*, 21, 319–327.

Pietroforte, R. and Gregori, T. (2006) Does volume follow share? The case of the Danish construction industry. *Construction Management and Economics*, 24(7), 711–715.

Rameezdeen, R., Zainudeen, N. and Ramachandra, T. (2005) Study of linkages between construction sector and other sectors of the Sri Lanakn economy. *Proceedings 15th International Input-Output Conference*, Renmin University in Beijing, China, June 27–July 1 (available online at http://www.iioa.org/Conference/15th-downable%20paper.htm – accessed 13 August 2007).

Rostow, W. W. (1956) The take-off into self-sustained growth. *Economic Journal*, 64, 25–48.

Rostow, W.W. (1960) *The Stages of Economic Growth: A Non-Communist Manifesto*, pp. 4–16. Cambridge: Cambridge University Press.

Rostow, W.W. (ed.) (1988) *The Economics of Take-Off into Sustained Growth*. Macmillan: London.

Ruddock, L. (1999) Optimising the Construction Sector: A Macroeconomic Appraisal – A case study of CEE countries. In: Ruddock, L. (ed) *Macroeconomic Issues, Models and Methodologies for the Construction Sector*, pp. 68–80. CIB Publication No. 240: Rotterdam, The Netherlands.

Ruddock, L. and Lopes, J. (2006) The construction sector and economic development: the 'Bon curve'. *Construction Management and Economics*, 24(7), 717–723.

Sen, A.K. (1983) Development: Which Way Now? *The Economic Journal*, 93, 745–762.

Song, Y., Liu, C. and Langston, C. (2005) Comparisons on linkages of construction and real estate sectors in OSCD countries. *QUT Conference Proceedings*. Brisbane: Australia

Song, Y., Liu, C. and Langston, C. (2006) Linkage measures of the construction sector using the hypothetical extraction method. *Construction Management and Economics*, 24(6), 579–589.

Strassmann, P. (1970) The Construction Sector in Economic Development. *Scottish Journal of Political Economy*, 17(3), 391–409.

Tan, W. (1993) Construction and economic development: the case of Singapore. *Habitat International*, 17(4), 75–87.

Tan, W. (2002) Construction and economic development in selected LDCs: past, present and future. *Construction Management and Economics*, 20, 593–599.

Turin, D.A. (1969) *Industrialization of Developing Countries: Problems and Prospects – Construction Industry*. UNIDO monograph No.2: New York.

Turin, D.A. (1973) *The Construction Industry: Its Economic Significance and Its Role in Development*. UCERG: University College London (also published as a UNIDO monograph, 1969).

Turin, D. (1978) Construction and development. *Habitat International*, 3(1), 33–45.

United Nations (1997) *International Recommendations for Construction Statistics. Statistics Division, Series M, No. 47, Rev. 1* (United Nations publication, Sales No. E.97.XVII.11).

United Nations Statistics Division, *National Accounts Main Aggregates Database*, http://unstats.un.org/unsd/snaama/status.asp (accessed 5 September 2007).

Wells, J. (1985) The role of construction in economic growth and development. *Habitat International* 9(1), 55–70.

Wells, J. (1986) *The Construction Industry in Developing Countries, Alternative Strategies for Development*. Croom Helm: London.

Chapter 3

Input–output techniques applied to construction

Tullio Gregori

Introduction

The famous approach of interindustrial relationships or input–output (IO) was formulated by Wassily Leontief in his celebrated paper *Quantitative Input and Output Relations in the Economic System of the United States* (1936) and then impressively developed by worldwide scholars.[1] Actually, the fundamental insight that an economy can be viewed as a set of interrelated actors was recognized by almost all classical economists. Cantillon (1755) and Quesnay (1758) are deemed the precursors, as they envisioned a society where social classes were interacting with each other. The former distinguished between landlords (the urban sector) and workers (the rural one), the first owning land and consuming luxuries, the second owning labour and consuming necessities, both exchanging commodities such that a balance was feasible. If workers' wages cannot buy necessities they will starve and die and, in turn, goods, which use labour as an input, will not be produced, so that landlords cannot buy their luxuries and they will be deprived too. Cantillon believed it possible to balance income and expenditure in a 'natural state' with 'natural prices', but land, that is exogenous and cannot be enlarged, constrains equilibrium ('The land is the source or matter from whence all wealth is produced', Cantillon, 1755: p. 2). He even calculated the amount of land needed to sustain workers' consumption putting the foundations of the so called 'iron law of wages'. However, he also recognized that 'it often happens that many things which have actually this intrinsic value {natural price} are not sold in the market according to that value: that will depend on the humours and fancies of men and on their consumption' (Cantillon, 1755: ch. 10).

Leontief himself acknowledged Quesnay's *Tableau Economique* as a source of inspiration for his IO model. Quesnay conceived a society comprising landlords, an agricultural and a manufacturing class. Then he showed how a bushel of corn or a tool could be produced using land, agricultural and manufactured goods with the famous *zig-zag* drawings in his tables. This approach can be related to the power series approximation in the Leontief model. Two main differences have been often overlooked. First, the *Tableau* is a monetary circuit, where firms pay each other for goods, as farmers buy crafts and artisans buy grain, while IO models rely on derived demand based on neoclassical optimizing firms. Furthermore IO analysis takes place in logical or local time, i.e. all transactions happen simultaneously, while a *Tableau* is a sequential representation where order and time both matter. Each node is not only a sector, but an industry at a particular point in time. One step in the production process requires another step that had taken place previously. Nodes must be considered in sequence so as to be able to decipher which transactions must be completed *before* other transactions are initiated. Hence, this approach allows us to describe how technical change propagates in an economy while this issue cannot be easily handled with fixed coefficient technologies.

Nonetheless Leontief (1941) was the very first economist who was not only constructing the *Tableau Economique* of the US from 1919 to 1929, but formally stating and solving a consistent model. His key insight still hinges on derived demand. In any economy, intermediate inputs are needed. Thus parts, as tyres, engines, windows and hubcaps, must be available to produce, say, a truck. In turn, tyres require rubber, valves, machinery and so on, while complex commodities such as engines are even more input demanding. Eventually more trucks are required to ship all these goods. Leontief's advancement was to introduce technical production relationships in the form of mathematical equations showing how much inputs are needed to produce any particular good. Such a model is now easily addressed and solved with the aid of a personal computer allowing us to figure out how much more of any commodity is needed.

This chapter is divided into three sections. First, IO accounting is introduced. While symmetric IO tables are usually adopted in empirical research they rely on 'make and use' rectangular tables which show production of domestic industries and purchases of commodities by firms and final users. Starting with the two basic accounting balances, this section illustrates what these rectangular tables represent and how they are linked with the symmetric one. Second, fundamentals of IO modelling are discussed. Most of the applied literature has straightforwardly used coefficients derived from compiled data without wondering about microeconomic foundations. In this section, the author briefly discusses production theory and how IO tables can be used to calibrate supply and demand functions. Third, a partial review of IO and related studies about the construction sector is provided, focusing on the interdependence between this industry and the rest of the economy presenting basic stylized facts. Finally, direction for further research is discussed.

Input–output accounting

IO analysis is based on the fundamental insight that an economy can be described by a system of linear equations, each one of which is showing flows of products from each producer to each consumer. These transactions can be arranged in a table which records flows for a particular period of time (usually a year) in monetary value terms. In the 'closed' Leontief model, all inputs are produced and all outputs exist merely to serve as inputs. Even labour, or any primary production factor, is a produced commodity, and a system is 'viable' if it satisfies its own needs, that is, there is no surplus or shortage (Pasinetti, 1977). In the standard or 'open' Leontief system we allow for discretionary consumption (final demand) and non-produced primary factors. Such a system can be described from either the supply or demand side. The latter adds industry (intermediate) to final demand, while the former shows production by firms. Let f_i depict final demand for the i-th commodity and u_{ij} the monetary value of the i-th commodity bought by j-th industry. The total use of commodities is given by the following system:

$$q_i = \sum_{j=1}^{n} u_{ij} + f_i \; i = 1,\ldots,m, \tag{1}$$

where m commodities can be demanded by n industries (or sectors) and final users. From (1) we can derive the 'use' or 'absorption' matrix $U = \{u_{ij}\}$ which records the commodities purchased and used by each industry as intermediate inputs to current production. These flows usually include both domestic and imported commodities and are valued at purchasers' prices. A standard 'use' table includes U and final demand too, i.e. vector f, or another matrix, that shows purchases by households, the government, investment (gross capital formation plus changes in inventories) and exports. The latter are in f.o.b. prices. Actually, (1) can also be a material balance, but compiled absorption tables are in monetary values with additional rows showing compensation of employees, taxes less subsidies on production, depreciation and operating surplus. These are payments for primary inputs required by industries as part of their production processes.

Theoretically, there is no reason to establish a one-to-one relationship between sectors and commodities. Industry accounts compile data assigning each establishment to a particular category according to its primary product. Joint production is widespread. The United Nation's International Standard Industrial Classification (ISIC) contains 17 major sections and 291 classes with four-digit coding, while the Central Product Classification (CPC) or other schemes, as CPA or PRODCOM, classify products based on the physical characteristics of goods or on the nature of the services rendered. The CPC detailed classification consists of 1,787 sub-classes (five-digit coding) and covers not only outputs of economic activities but land or licence transactions as well. These assets, as patents or copyrights, are not regarded as goods and services in the System of National Accounts (SNA). On the contrary standard IO analysis assumes that each industry is producing a single commodity and $n = m$. However, this is not the case for many activities. For example, construction may be engaged not only in building activities but engineering services too. Such a simplifying assumption is not needed if we know which commodities are produced by each industry. Hence sectoral total output is the sum of its deliveries of any commodity measured at basic prices[2] (i.e. without trade and transport margins and net taxes on products):

$$g_i = \sum_{j=1}^{m} v_{ij} \, i = 1, \ldots, n. \tag{2}$$

This system of equations can be arranged in a table called the 'make' or 'supply' matrix. Here, we can read the value of each commodity produced by each industry. Domestic industries are classified according to the principal product. They are listed at the beginning of the rows, while commodities are named at the head of the columns. Row entries show the product mix of each industry as they depict its primary and secondary products. In a square matrix, the value of the primary product of an industry is shown in the diagonal cell while secondary products are shown in the off-diagonal entries along the row. We can easily assess the extent to which each sector specializes in the production of its primary output and the range of secondary activities. Moreover, entries in a column represent the value of production by each industry of the commodity named at the head of the column. Hence, each column shows how commodity production originated from domestic industries. Adding commodity taxes and subsidies, we get domestic supply, while total supply includes imported goods at c.i.f. prices. Finally, by allocating trade and transport margins from the trade and transport sectors to the traded and transported goods and by adding commodity net taxes, output in purchasers' prices is obtained. The complete set of matrices is represented in Figure 3.1, which is based on Eurostat (1996), where matrix dimensions have been previously defined as **V** is the $(n \times m)$ make matrix, **U** the $(m \times n)$ use matrix, q is the $(m \times 1)$ commodity (or product) gross output vector, g the $(n \times 1)$ industry total output vector and u the $(1 \times n)$ value added vector.

For sake of simplicity, we assume that final demand and value added are vectors whose dimensions are respectively $(m, 1)$ and $(1, n)$.[3] Given these definitions we can easily derive the commodity by industry direct requirement matrix:

$$\mathbf{B} = \mathbf{U}(\hat{g})^{-1}, \tag{3}$$

which entry $b_{ij} = u_{ij}/g_j$ shows the amount of the i-th commodity required to produce one unit of the j-th industry. Hence, postmultiplying by $\hat{g}$ and using (1), we can state:

$$q = \mathbf{B}g + f. \tag{4}$$

	Products	Industries		Final demand			Total
		Use	Private	Government	Gross capital	Exports	Total use of
Products		Matrix	consumption	consumption	formation		products
Industries	Make matrix						Total domestic output
Taxes and subsidies	Product net taxes						
Value added		Value added					
Imports	Imported products						
Total	Total supply of products	Total domestic output					

Figure 3.1 Make and use matrices.

Total commodity output q is equal to intermediate commodity production, that is commodity by industry direct requirements times total industry output, plus commodity deliveries to final demand. If we divide the Make matrix by sectoral output:

$$C = V'(\hat{g})^{-1}, \tag{5}$$

we can assess how much of the total production in any industry is attributable to the production of the j-th commodity as $c_{ij} = v_{ij}/g_i$ shows industry output proportions. Finally, if we divide V by commodity total production:

$$D = V(\hat{q})^{-1}, \tag{6}$$

we get commodity quotas produced by each industry as $d_{ij} = v_{ij}/q_j$, which measures the fraction of total production of the j-th commodity produced by the i-th industry.

These matrices provide a very detailed picture of all economic activities as production, consumption, accumulation and trade. The use matrix gives information on the uses of goods and services, and on cost structures of the industries. Its row total is total commodity output (regardless of which industry contributed to that output) and column total is total industry output (regardless of what commodity was produced). Column totals of the make matrix are equal to row totals of the full use matrix, equivalent to commodity outputs. On the other hand, row totals of the make matrix are equal to column totals of the full use matrix, equivalent to the industry outputs.

From these matrices, a symmetric input–output (SIO) table is often derived in which the same classification is used in both rows and columns. Theoretically, either an industry-by-industry or a commodity-by-commodity table can be constructed as we can deem either institutionally defined industries or homogeneous units characterized by producing certain commodity groups (UN, 1999). In the former, product technologies are depicted in columns of the SIO matrix, while rows represent the distribution of products to intermediate and final users. The latter considers industries as groups of establishments or enterprises. The dilemma about this choice is not easily solvable but the commodity-by-commodity table is more popular and recommended by UN and Eurostat (UN, 1999; Eurostat, 2002). In this case, secondary products must be transferred, so that they are treated as additions into the activities for which they are principal and removed from the activities in which they are produced. Inputs associated with secondary outputs must be removed from the industry in which that secondary output actually takes place to the activity to which characteristically belongs. The mathematical methods used to transfer outputs and associated inputs hinge on two conjectures: either the industry or the commodity technology assumption. The former takes for granted that all products (whether principal or secondary) produced by an industry have the same input structure, while the commodity technology hypothesis assumes that a product has the same input structure in whichever industry it is produced. The industry technology assumption allows us to figure out the direct coefficient matrix in the commodity-by-commodity IO model as $\mathbf{A}_{I,cc} = \mathbf{BD}$. Actually from (4) we get:

$$q = \mathbf{B}g + f = \mathbf{BD}q + f, \tag{7}$$

since (6) can be stated as $g = \mathbf{D}(\hat{q})i = \mathbf{D}q$ where i is a $(n \times 1)$ unit vector. Even the industry-by-industry model can be derived as:

$$\mathbf{D}q = g = \mathbf{B}g + \mathbf{D}f. \tag{8}$$

Unfortunately, the industry technology assumption that has been recommended by the 1968 SNA is no longer acceptable as it breaks the fundamental economic rule that products with different prices at a given moment must reflect different costs or technologies (UN, 1999). The commodity technology assumption makes more economic sense, but it requires the direct coefficient matrix to be $\mathbf{A}_{C,cc} = \mathbf{BC}^{-1}$ (Miller and Blair, 1985). Hence $\mathbf{V}$ and $\mathbf{C}$ must be square matrices since the latter has to be inverted. It's quite unrealistic that the number of products is equal to the number of industries. Moreover its automatic mathematical derivation can produce unacceptable results as net inputs must be figured out and negative technical coefficients can emerge (UN, 1999). Hence more sophisticated techniques, as minimization of variances or entropy maximization subject to constraints, have been devised so that material balance is fulfilled[4] (Golan et al., 1994; Robinson et al., 2000). Actually we can distinguish between two distinct forms of equilibrium. If T is the interindustry transaction flow matrix, then total output by industry is given by the sum of final and interindustry demands:

$$g = Ti + f \qquad (9)$$

and by the supply of primary inputs and interindustry supplies from all sectors forming the national economy:

$$g' = w + i'T. \qquad (10)$$

As for the use matrix, (9) can be in either physical or value units (Gold, 1993; Weisz and Duchin, 2006) while (10) makes sense in monetary flows only. Systems (9)–(10) provide the analytical framework for most of economic modelling and construction analyses. These issues are tackled in the following sections.

Input–output models

IO tables are powerful tools for compiling production accounts and have been part of the SNA since 1968, even if the technology description can be in physical terms too (Katterl and Kratena, 1990; Stahmer et al., 1998; Pedersen, 1999). Equations (9)–(10) are accounting equations with little economic foundations because, ex post, they must, by definition, hold. Nonetheless, their inspection provides useful insights about the working of an economy and its sectors in a particular year. Several measures have been proposed to quantify the importance of each industry or to characterize construction. Let us first consider the economic indicators formed from raw data that are theoretically free. These are, among others, construction output (or value added) over national output (value added), construction final demand to total demand (or total final demand), construction intermediate input and output to national output (Bon, 1988). All these indices can show its role in any macroeconomic system, but do not describe its technology in terms of disaggregated inputs and deliveries to other sectors. Actually, IO tables are very useful to address these issues as it is well known that construction value added and employment are rather small in comparison with other sectors, while intermediate utilizations are quite large. Total resource use deserves special attention because of its assembly character, and sectoral interdependence is significant. For this reason, researchers have provided extensive comparative analyses over time and between countries. However, almost all the applied literature forgets that technological structural relationships would perhaps be measured more accurately in physical units while most of compiled tables report flows in value terms. This is not an issue when sectors are compared in a point in time (or period) or the Leontief model is true and the law of unique price holds. To address these issues, let us consider a hypothetical two-sector economy (construction (Const) and the rest of the economy (RoE)) adopted by Miller and Blair (1985: p.355) starting from the table in physical terms (Table 3.1). From these flows a similar matrix in monetary units can be immediately derived.

Here, labour is the only primary factor, whose (exogenous) wage is equal to 10 (say €), and (endogenous) commodity prices are €2 for RoE and €5 for construction. Actually intermediate transactions and final demand figures in money values can be worked out from the dual or the IO price model. Let the input or technical coefficient matrix be:

$$A = X(\hat{x})^{-1}, \tag{11}$$

$$S = T(\hat{p}\hat{x})^{-1}, \tag{12}$$

where X is the physical flow matrix and T the value one. It can be shown that profit maximizing industries endowed with Leontief production functions (i.e. with zero substitution elasticities) will choose outputs and prices according to:

$$x = (I - A)^{-1} y, \tag{13}$$

$$p = (I - A')^{-1} v, \tag{14}$$

where v is the $(n \times 1)$ value added coefficient vector or, in our example, labour per unit of output. System (13) is an equilibrium material balance with exogenous final demand, while industries are price setting due to constant returns to scale implied by Leontief production functions (Varian, 1992). Commodity prices are dictated by (capital and) labour contents alone and this price model[5] can be used to assess inflation transmission (Melvin, 1979). Matrix inverses in (13)–(14) are useful for impact analysis, as their entries, called multipliers, show changes in outputs and prices due to final demand or value-added variations.

Moreover, from (14) we can compute relative prices. In our hypothetical example, commodity RoE is valued 0.2 of labour price while construction 0.5.

Table 3.1 Transaction matrices

Physical flows					Value-based flows				
	RoE	Const	y	X		RoE	Const	y	x
RoE	75	250	175	500	RoE	150	500	350	1000
Constr.	40	20	340	400	Constr.	200	100	1700	2000
Labour	65	140			Labour	650	1400		

In monetary terms, the former is €2 and the latter €5 when the wage is €10. All consumers (intermediate and final) pay the very same prices for these goods (*unique price law*) and the IO table in monetary terms can be immediately figure out as in Table 3.1 where balance equation (10) can be checked.[6] Technical coefficient matrices can be immediately derived as in Table 3.2, and we can check the following relationship:

$$S = \hat{p}A(\hat{p})^{-1}. \tag{15}$$

In the value-based flow matrix, column sums add to unity by definition[7] as total outlays is equal to total output value. Hence, columns show expenditure shares and each entry is the euro's worth of an input per euro's worth of an industry output. In our example, construction is paying labour a lot (70% of its production value) while intra-industry flow is quite small (5% of total factor payments). Inverse matrices can be immediately calculated and compared, but, up to now, only the physical one has a proper microeconomic foundation.

First of all, we can notice striking differences in off diagonal entries. Let us consider the second element in the first row. This entry is more than twice as large in the physical than in the value inverse. Actually economic interpretation of (13) is straightforward: a unit increase in (physical) construction final demand will induce a 0.825 expansion in RoE (physical) production. We can apply the same logic and state that a €1 increase in final demand will increase RoE output value by only €0.33. Prices matter as the relative RoE-Constr. price is 2/5, and our statement is sound.[8] However, we have implicitly assumed that relative prices are not changing when we increase final demand and output. This is a real issue since engineering data are rarely available and almost all IO analyses have accepted the value matrices. The standard solution is to normalize prices so that the price vector is the unit vector i. This is accomplished taking value added coefficients per unit (euro's worth) of output. In our example $v' = (0.65, 07)$ as shown in the last row of Table 3.2: applying (14), we get a unit price vector and shares are equal to 'physical' input coefficients in the base year data set. This approach is questionable if we know the true technology but is otherwise needed. It can be claimed that if, in the Leontief model, the true value input coefficients s_{ij} are as stable as a_{ij}

Table 3.2 Technical coefficient matrices and Leontief inverses

Physical flows					Value-based flows				
	RoE	Constr.	Inverse	Coeff		RoE	Constr.	Inverse	Coeff
RoE	0.150	0.625	1.254	0.825	RoE	0.150	0.250	1.254	0.330
Const	0.080	0.050	0.106	1.122	Const	0.200	0.050	0.264	1.122
Labour	0.130	0.350			Labour	0.650	0.700		

when value added is not changing, as dictated by (14). Economic history has demonstrated relative prices and value-added inputs have moved a lot in the past, and coefficient or multiplier comparisons have little meaning over time as each round we normalize prices ignoring actual inflation. Changes in technical coefficients may reflect changes in the cost structure rather than in the production structure or vice versa. The quantity effect is not separated from the price effect not allowing a proper understanding of the construction technology. Let us assume the following true technical change. Construction is now input saving as it needs less RoE and labour inputs. New coefficients are shown in Table 3.3 with value shares derived from actual flow data.[9] According to value-based flows, there is almost no change in technical coefficients as the price increase in the RoE commodity offsets most of the input requirement decrease per unit of output.[10] Furthermore RoE output multipliers are the same as in Table 3.2. Hence, a quite large structural change (15 per cent less of RoE and 8 per cent less of labour inputs) is practically unnoticed. Hence, for a proper understanding of the changes in the production structure over time, it is necessary to use the IO table in constant prices.

Unfortunately, tables cannot be purely double deflated as in (15) because results can be distorted and suffer aggregation bias too (Sevaldson, 1976; Wolf, 1994). When statistics bureaus do not provide tables in constant prices, alternative easy procedures are available (Folloni and Miglierina, 1994; Dietzenbacker and Hoen, 1998). However, price influence has not been addressed in construction IO studies and even matrices in constant prices have been rarely adopted (Pietroforte and Gregori, 2003). This calls for more accurate investigation of the decreasing importance of the industry itself and the progressive transformation of its technologies, with decreasing manufacturing inputs and increasing service input. It would be interesting to know how much is due to price and quantity changes.

Furthermore, similar difficulties arise when input and inverse matrices of

Table 3.3 Effects of technical change in the construction sector

Physical flows					Value-based flows				
	RoE	Const	y	X		RoE	Const	y	x
RoE	75	212	175	462	RoE	162	457	378	997
Const	40	20	340	400	Const	185	92	1572	1850
Labour	65	130			Labour	650	1300		
	Direct		Inverse			Direct		Inverse	
RoE	0.163	0.525	1.267	0.707	RoE	0.162	0.247	1.267	0.330
Const	0.087	0.050	0.115	1.117	Const	0.186	0.050	0.247	1.117
Labour	0.141	0.325			Labour	0.652	0.703		

different countries are compared in an effort to establish similarities or differences in production techniques. If commodity (or primary factors) relative prices differ a lot these assessments can likely reveal little about differences in technologies or even produce meaningless results. Ideally, purchasing power parity (PPP) should be embraced (Kravis, 1984; Heston and Summers, 1996), but tables in PPP prices are not published and only scattered data related to construction are available (Heston et al., 2006).

Another solution is to switch to a different model. Klein (1952–53) and Morishima (1956–57) proposed the use of monetary IO tables to calibrate a log linear model. As in the Leontief model, it can be shown that profit maximizing industries endowed with Cobb Douglas production functions (i.e. with constant substitution elasticities) will choose outputs and prices according to:

$$\ln x = (I - S)^{-1} \ln y, \qquad (16)$$

$$\ln p = (I - S')^{-1} \ln v, \qquad (17)$$

where the value-based input matrix has been adopted. 'It has been suggested that input-output specialists reason in terms of Cobb-Douglas model rather than fixed coefficient production functions' (Moses, 1974: 13) as they embrace monetary values. Multipliers in (16)–(17) are output and price elasticities. This model is as useful as the Leontief one for short-run analysis since, over the long haul, technical change can be relevant. Then IO coefficients in physical terms can vary not only because of productivity but relative prices changes too and input ratios derived from compiled tables in actual prices should be more stable than those calculated from deflated ones. Unfortunately applied studies are rather inconclusive and disentangling these sources of structural change is far from easy (Tilanus and Rey, 1964; Bezdek and Wendling, 1976; Harmston and Chow, 1980).

Whatever the production function, prices do not determine either input or share choices and production depends on final demand alone while, in turn, value-added composition does not affect output. Two further implicit assumptions cannot be neglected: final demand and primary inputs are exogenous or inelastic with respect to prices and all input supplies are perfectly elastic, even labour and capital. The last hypothesis is rather demanding as it requires excess capacity and no wage or interest rate response to larger production (Moses, 1974). To acknowledge these relationships we must resort to CGE models, where all demands and supplies interact. In the literature it has been suggested the supply model developed by Ghosh (1958) and Augustinovics (1970) can give some insights on this issue. This is an alternative approach to the Leontief model that is still based on the transaction matrix T but with supply that determines production. The supply driven or Ghosh model assumes a fixed allocation of outputs over sectors, known as sales or allocation coefficients:

$$G = (\hat{x})^{-1}X, \tag{18}$$

that is formed by dividing the flow from one industry to another by the total output of the former industry. It is interesting to notice that, without price discrimination, matrix G is made by pure numbers even if X is substituted by the flow matrix T as it records output deliveries to all the sectors i.e. $g_{ij} = t_{ij}/(p_i x_i) = (p_i x_{ij})/(p_i x_i) = x_{ij}/x_i$. Ghosh claims that, in monopolistic markets or when there are shortages of resources, output coefficients are even more stable than input coefficients. Then, it is customary to apply the identity (10) and derive its well known solution which relates exogenous primary factors to gross output. Unfortunately, (10) makes sense in monetary units only. Hence, gross output is measured in value too, while, for a long time the supply model was believed to be a quantity model (Oosterhaven, 1988: p. 205). This false interpretation was suggested by Augustinovics and Ghosh and widely applied in impact studies (Jones, 1976; Giarratani, 1976; Bon, 1984, 1988) and construction analyses too (Bon, 1988; Bon and Pietroforte, 1990). This approach has been criticized by several scholars, since the traditional Leontief model is grounded on well established production theory, whereas the supply-driven model seems to lack proper theoretical foundations and, when used together with (13), yields implausible results (Cella, 1988; Oosterhaven 1988, 1989). Let us assume an increase in value added, because there is a larger availability of some factors such as labor, in sector j. Then the (quantity) supply-driven model suggests that there is an increase in (real) output over all sectors. Hence, in any sector other than the jth, the production is augmented without any change in value added. This is obviously at odds with traditional Leontief production functions, which imply a proportional increase in all factors even primary. Dietzenbacher (1997) suggests a new interpretation. He shows that the supply-driven model yields exactly the same results as the Leontief price model (14) with fixed quantities and proposes to define the supply driven model as the '*Ghosh price model*' that can be used for cost push exercises. Using (10) and G we get:

$$(\hat{x}p)' = w + i'T = w + (\hat{x}p)'G \tag{19}$$

whose solution is well known:

$$(\hat{x}p)' = w\,(I - G)^{-1}. \tag{20}$$

The working of the model rests on the hypothesis that all physical quantities remain unchanged as final demand is fixed and the only driving forces are value-added changes. These can be either real or nominal or both. Furthermore there is a dual to (20) that is the Ghosh quantity solution which gives real quantity changes from quantity variations in final demand. Nonetheless, these add little to Leontief-type models (De Mesnard, 2007) and the 'usefulness of the Ghosh-type models is to be sought in providing a better (i.e. simpler or more straightforward) interpretation of multipliers, forward linkages and the effects of cost-pushes' (Dietzenbacher, 1997: p. 631).

Construction in input–output analysis

According to a vast literature, a logistic curve can be used to explain urban population share as an economy grows (Berry, 1973). At earlier stages of development, the urban to total population ratio and building activity are first growing at an increasing rate and eventually at a decreasing one. As labour and value added move to services, construction share is declining. This relationship implies a bell-shaped share of construction in GNP analogous to the pattern of relative deindustrialization with a declining share of physical assets in investment (Kuznets, 1968; Maddison, 1987). Several studies confirm this GNP inverted U-shaped curve (Bon, 1988, 1991; Bon and Minami, 1986; Bon and Pietroforte, 1990; Pietroforte and Gregori, 2003) while it is questioned if it holds for its volume as well (Pietroforte and Gregori, 2006; Ruddock and Lopes, 2006; Mehemet and Yorucu, 2008). IO tables can provide useful insights as to whether this sector is a growth engine in developed countries too, as it can generate further output due to intersectoral linkages. In IO tables construction is depicted by at least one row and column, but it can be divided into 'new construction' and 'maintenance and repair' (M&R). The former includes private and public new buildings, addition and alterations that increase the stock of constructed facilities. M&R comprises restoration and upkeeping of existing capital stock performed also on own account. Columns in the SIO or use matrices illustrate input profiles. Bon and Pietroforte (1993) show how service inputs became dominant in new construction in the 1970s in the US, while manufacturing and petroleum usage was shrinking. As concerns M&R, wholesale and retail are the key direct suppliers and almost all input contributions did not change in the 1960s and 1970s. The only exception is petroleum again. This finding is not surprising, as US IO tables are not deflated and oil shocks matter. Pietroforte and Gregori (2003) address this issue with the OECD dataset in actual and constant prices. Unfortunately, these SIO tables do not distinguish between new construction and M&R. They find, on average, an increase in service input and a decrease in manufacturing requirements in both data sets. While the upward trend in tertiary is a common feature with some significant accelerations (as in Denmark, France and Germany), manufacturing patterns are quite different. This calls for country studies that are part of a growing literature (Bon and Pietroforte, 1990; Bon and Yashiro, 1996; Chan, 2001; Lean, 2001; Su et al., 2003; Rameezdeen et al., 2005; Wu and Zhang, 2005; Song and Liu, 2006; Song et al., 2007; Mehemet and Yorucu, 2008).

Construction deliveries are analysed by means of the allocation matrix **G**. Its entries are called direct forward linkages as they show intermediate to total output ratios. They should be price-free indicators, but pure double deflation is rarely used and price discrimination can exist. Comparative studies with actual/

constant tables are needed to shed light on this issue. Some studies exhibit very stable shares over quite long timespans with country differences (Bon, 1988; Pietroforte and Bon, 1995, 1999; Bon and Yashiro, 1996). Other empirical analyses present large time variability or mixed results (Bon, 1991; Bon et al., 1999; Pietroforte and Gregori, 2003; Rameezdeen et al., 2005).

The quantity Leontief model is widely applied to figure out total backward output multipliers, that is total impact of changes in final demand on sectoral output. It is customary to measure the total effect of a unit change in final demand for the goods and services of the building industry on the output of 'all' industries, that is all the entries in the column that refer to construction in the Leontief inverse are summed up. This makes sense if the reference model is the standard one, i.e. (13), and either a physical flow matrix is known or all the prices are set to unity.[11] As above, time series comparisons require IO tables in constant prices otherwise these exercises are meaningful only if (relative) prices are stable. There is plenty of evidence about the positive correlation between relative price variability and inflation, even between countries, so that this implicit assumption is not tenable (Fisher, 1982; Hartman, 1991; Silver and Ioannidis, 2001).

The overall 'push effect' of the building industry on the economy as a whole is usually calculated with the total forward linkage indicators or output multipliers derived from the Ghosh inverse. These multipliers should measure the effect of a monetary unit change in primary inputs available to an industry on the output of other industries. Yet the endogenous variable in the supply model is nominal output with fixed quantities. Hence, all published results should be read as pure cost push exercises and compared with sectoral price variability analyses.

Conclusions

IO tables provide detailed information on production and consumption activities in an economy by recording all the transactions between producers and consumers. They form an important part of the System of National Accounts harmonized by the United Nations (SNA 93) and the European Union (Eurostat, 1996) and are made up of a set of tables. The most well known and widely compiled are the make, the use and the SIO tables.[12] They provide a framework to analyse the uses of domestically produced and imported goods and services and allow us to assess direct, indirect and induced changes on the whole economy by a change in exogenous variables. While direct requirements and sales can be easily figured out from compiled tables, standard IO models rely on explicit and implicit assumptions. A fundamental one is that, ex ante or in any period, interindustry quantity flows depend exclusively on total output for that same time period. This is to assume a Leontief production function that constrains profit-maximizing behaviour. The optimal solution is the well known quantity model whose inverse has been widely applied but a dual price model too. Unfortunately parameter calibration requires quantity data or to fix prices to unity or to resort to the Klein-Morishima model with Cobb-Douglas production functions. In any case the so-called supply or Ghosh model cannot be accepted as a quantity model but as an equivalent price model for cost push exercises.

Using this framework we can read most empirical research. A large share of applied literature deals with IO in actual prices without questioning if inflation and relative price variability (or country price variability) were issues in the period under scrutiny. Results must be appraised with caution particularly in the 1970s and 1980s, when oil shocks and price upsurges were common phenomena. Actually, large total forward multipliers indicate commodity price can change a lot when value added is not stable.

Nonetheless, a few stylized facts are well known. Construction seems to follow the economic destiny of manufacturing, its primary supplier and key source of its demand, whose bell-shaped pattern of industrialization and deindustrialization is acknowledged. There has been a shift of importance of some intermediate inputs away from manufacturing towards service inputs and, in more advanced countries, a change in the composition of primary inputs with relative decreasing importance of manpower. Hence, better data and further enquiries, even with new tools (Liu and Song, 2005; Chiang et al., 2006; Gregori and Pietroforte, 2006; Song et al., 2006a, 2006b), are needed to fruitfully assess changes in interdependence. However, the field appears to be healthy and growing.

Notes

1 The huge collection edited by Kurz, Dietzenbacher and Lager (1998), totalling more than 1500 pages, is just a tiny fraction of the impressive empirical and theoretical literature on this subject as Taskier's bibliography, published by the United Nations in the 1960s, already contained 10,000 works.
2 Here a material balance is nonsense.
3 Actual tables are obviously more detailed.
4 Row and column totals have to be identical for each product.
5 A different model with capital and workers' consumption matrices is presented in Sekerka et al. (1970).
6 Zero profit is assured by constant returns.
7 Column sums are meaningless for physical flows as elements are measured in different units.
8 Of course 0.33 = 0.825*2/5.
9 The price system (14) has been applied again. The construction price has dropped to €4.62 and RoE price increases to €2.16, while the labour wage is still fixed to €10.
10 With the same physical final demand, output is lower in money value.
11 If we adopt the Klein-Morishima approach, inverse entries are elasticities whose sum is meaningless.
12 The European Union has made their compilation mandatory for its member countries.

References

Augustinovics, M. (1970) Methods of international and intertemporal comparisons of structure. In Carter, A.P. and Brody, A. (eds) *Contributions to Input-Output Analysis*, pp. 249–269. North Holland: Amsterdam.

Berry, B.J.L. (1973) *The Human Consequences of Urbanization: Divergent Paths in the Urban Experience of the Twentieth Century*. MacMillan: London.

Bezdek, R.H. and Wendling, R.W. (1976) Current and constant dollar input-output forecasts for the U.S. economy. *Journal of the American Statistical Association*, 71, 543–551.

Bon, R. (1984) Comparative stability analysis of multiregional input-output models: column, row and Leontief-Strout gravity coefficients models. *The Quarterly Journal of Economics*, 99, 791–815.

Bon, R. (1988) Direct and indirect resource utilization by the construction sector: the case of the USA since World War II. *Habitat International*, 12, 49–74.

Bon, R. (1991) What do we mean by building technology? *Habitat International*, 15, 3–26.

Bon, R. (1992) The future of international construction: secular patterns of growth and decline. *Habitat International*, 16, 119–128.

Bon, R. (2000) *Economic Structure and Maturity*. Ashgate Publishing Ltd: Aldershot.

Bon, R. and Minami, K. (1986) The role of construction in the national economy: a comparison of the fundamental structure of the US and Japanese input-output tables since the World War II. *Habitat International*, 10, 93–99.

Bon, R. and Pietroforte, R. (1990) Historical comparisons of construction sectors in the United States, Japan, Italy and Finland using input-output tables. *Construction Management and Economics*, 8, 233–247.

Bon, R. and Pietroforte, R. (1993) New construction versus maintenance and repair construction technology in the US since World War II. *Construction Management and Economics*, 11, 151–162.

Bon, R. and Yashiro, T. (1996) Some new evidence of old trends: Japanese construction 1960–90. *Construction Management and Economics*, 14, 319–323.

Bon, R., Birgonul, T. and Ozdogan, I. (1999) An input-output analysis of the Turkish construction sector, 1973–1990: a note. *Construction Management and Economics*, 17, 543–551.

Cantillon, R. (1755) *Essai sur la Nature du Commerce en Général* (Essay on the Nature of Trade in General), edited and translated by H. Higgs (1931), McMillan, London.

Cella, G. (1988) The supply side approaches to input-output analysis: an assessment. *Ricerche Economiche*, XLII, 433–451.

Chan, S.L. (2001) Empirical tests to discern linkages between construction and other economic sectors in Singapore. *Construction Management and Economics*, 19, 355–363.

Chiang, Y.-H., Cheng, E. and Tang, B.-S. (2006) Examining repercussions of consumptions and inputs placed on the construction sector by use of I-O tables and DEA. *Building and Environment*, 41, 1–11.

De Mesnard, L. (2007) *About the Ghosh model: Clarification*. LEG, Economy Series, Working Paper No. 2007–06.

Dietzenbacher, E. (1997) In vindication of the Ghosh model: a reinterpretation as a price model. *Journal of Regional Science*, 37, 629–651.

Dietzenbacher, E. and Hoen, A.R. (1998) Deflation of input-output tables from the user's point of view: a heuristic approach. *Review of Income and Wealth*, **44**, 111–122.

Eurostat (1996) *European System of Accounts (ESA95)*. EU.

Eurostat (2001) *Eurostat Input-Output Manual*. 1st draft.

Eurostat (2002) *Symmetric Input-Output Tables*. Workshop on Compilation and Transmission of Tables in the Framework of Input-Output System in ESA95, Luxemburg, doc. B.6.

Fisher, I. (1982) Relative price variability and inflation in the United States and Germany. *European Economic Review*, **18**, 171–196.

Folloni, G. and Miglierina, C. (1994) Hypothesis of price formation in input-output tables. *Economic System Research*, **6**, 249–264.

Ghosh, A. (1958) Input-output approach to an allocation system. *Economica*, **25**, 58–64.

Giarratani, F. (1976) Application of an interindustry supply model to energy issues. *Environment and Planning A*, **8**, 447–454.

Golan, A., Judge, G. and Robinson, S. (1994) Recovering information from incomplete or partial multisectoral economic data. *Review of Economics and Statistics*, **76**, 541–549.

Gold, M. (1993) Alternative approaches of physical input-output analysis to estimate primary material inputs of production and consumption activities. *Economic System Research*, **16**, 301–310.

Gregori, T. and Pietroforte, R. (2006) How much labor in construction? In: Pietroforte, R., De Angelis, E. and Polverino, F. (eds) *Construction in the XXI Century: Local and Global Challenges*, pp. 128–129. Edizioni Scientifiche Italiane.

Guccione, A. (1986) The input-output measurement of interindustry linkages: a comment. *Oxford Bulletin of Economics and Statistics*, **48**, 373–377.

Harmston, F.K. and Chow, W.S. (1980) A test of current versus constant dollar input-output multipliers: the Missouri case. *The Review of Economics and Statistics*, **62**, 127–130.

Hartman, R. (1991) Relative price variability and inflation. *Journal of Money, Credit & Banking*, **23**, 185–205.

Heston, A. and Summers, R. (1996) International price and quantity comparisons: potentials and pitfalls. *American Economic Review*, **86**, 20–24.

Heston, A., Summers, R. and Aten, B. (2006) *Penn World Table Version 6.2*. Center for International Comparisons of Production, Income and Prices: University of Pennsylvania.

Jones, L.P. (1976) The measurement of Hirschmanian linkages. *The Quarterly Journal of Economics*, **90**, 323–333.

Katterl, A. and Kratena, K. (1990) *Reale Input-Output Tabelle und ökologisher Kreislauf*. Physica: Heidelberg.

Klein, L.R. (1952–1953) On the interpretation of Professor Leontief's system. *Review of Economic Studies*, **20**, 131–136.

Kravis, I.B. (1984) Comparative studies of national income and prices. *Journal of Economic Literature*, **22**, 1–39.

Kurz, H.D., Dietzenbacher, E. and Lager, C. (1998) *Input–Output Analysis*. Edward Elgar: Cheltenham.

Kuznets, S. (1968) *Toward a Theory of Economic Growth*. Norton: New York.

Lean, S.C. (2001) Empirical tests to discern linkages between constructions and other economic sectors in Singapore. *Construction Management and Economics*, **19**, 355–363.

Leontief, W. (1936) Quantitative input and output relations in the economic system of the United States. *Review of Economics and Statistics*, **18**, 105–125.

Leontief, W. (1941) *The Structure of the American Economy, 1919–1929: an Empirical Application of Equilibrium Analysis.* Harvard University Press: Cambridge, Mass.

Leontief, W. (1986) *Input-Output Economics.* Oxford University Press: New York.

Liu, C. and Song, Y. (2005) Multifactor productivity measures of construction sectors using OECD input-output database. *Journal of Construction Research*, **6**, 209–222.

Liu, C., Song, Y. and Langston, C. (2005) Economic indicator comparisons of multinational real estate sectors using the OECD input-output database. *The International Journal of Construction Management*, **4**, 59–75.

Maddison, A. (1987) Growth and slow down in advanced capitalist economies. *Journal of Economic Literature*, **25**, 649–98.

Mehemet, O. and Yorucu, V. (2008) Explosive construction in a micro-state: environmental limit and the bon curve: evidence from North Cyprus, *Construction, Management and Economics*, **26**, 79–88.

Melvin, J.R. (1979) Short-run price effects of the corporate income tax and implications for international trade. *American Economic Review*, **69**, 765–774.

Miller, R.E. and Blair, P.D. (1985) *Input-Output Analysis: Foundations and Extensions.* Prentice-Hall: Englewood Cliffs.

Morishima, M. (1956–1957) A comment on Dr. Klein's interpretation of Leontief's system. *Review of Economic Studies*, **24**, 65–68.

Moses, L.N. (1974) Outputs and prices in interindustry models. *Papers of Regional Science Association*, **32**, 7–18.

Oosterhaven, J. (1988) On the plausibility of the supply-driven input-output model. *Journal of Regional Science*, **28**, 203–217.

Oosterhaven, J. (1989) The supply-driven input-output model: a new interpretation but still implausible. *Journal of Regional Science*, **29**, 459–465.

Pasinetti, L.L. (1977) *Lectures on the Theory of Production.* MacMillan: London.

Pedersen, O.G. (1999) *Physical input-output tables for Denmark products and materials, 1990. Air emissions 1990–92.* Statistics: Denmark.

Pietroforte, R. and Bon, R. (1995) An input-output analysis of the Italian construction sector 1959–1988. *Construction Management and Economics*, **13**, 253–262.

Pietroforte, R. and Bon, R. (1999) The Italian residential construction sector: an input–output historical analysis. *Construction Management and Economics*, **17**, 297–303.

Pietroforte, R. and Gregori, T. (2003) An input-output analysis of the construction sector in highly developed economies. *Construction Management and Economics*, **21**, 319–327.

Pietroforte, R. and Gregori, T. (2006) Does volume follow share? The case of the Danish construction industry. *Construction Management and Economics*, **24**, 711–715.

Polenske, K. and Sivitanides, P. (1990) Linkage in the construction sector. *Annals of Regional Science*, **24**, 147–161.

Pressman, S. (1994) *Quesnay's Tableau Economique: a Critique and Reassessment.* August M. Kelley: Fairfield, NJ.

Quesnay, F. (1758) *Tableau Economique* (The Economic Table) edited and translated by University Press of the Pacific (2004).

Rameezdeen, R., Zainudeen, N. and Ramachandra, T. (2005) Study of linkages between construction sector and other sectors of the Sri-Lankan economy. *Proceedings of the 15th International Conference on Input-Output Techniques*, June 25-July 1, 2005, Renmin University of China, Beijing, China.

Robinson, S., Cattaneo, A. and El-Said, M. (2000) Updating and estimating a social accounting matrix using cross entropy methods. *Economic System Research*, 13, 47–64.

Ruddock, L. and Lopes, J. (2006) The construction sector and economic development: the 'Bon Curve'. *Construction Management and Economics*, 24, 717–723.

Sekerka, B., Kyn, O. and Hejl, L. (1970) Price systems computable from input-output coefficents. In: Carter, A.P. and Brody, A. (eds) *Contributions to Input Output Analysis*, pp. 183–203. North Holland: Amsterdam.

Sevaldson, P. (1976) Price changes as causes of variations in input-output coefficients. In: Polenske, K.R and Skolka, J.V. (eds) *Advances in Input-Output Analysis*, pp. 207–237. Ballinger: Cambridge.

Silver, M. and Ioannidis, J. (2001) Intercountry differences in the relationship between relative price variability and average prices. *Journal of Political Economy*, 109, 355–374.

Song, Y. and Liu, C. (2005) Economic performance analysis of the Australian property sector in the 1990s using the input-output tables. *Journal of the Pacific Rim Real Estate Society*, 11, 412–425.

Song, Y. and Liu, C. (2006) The Australian construction linkages in the 1990s. *Architectural Science Review*, 49, 408–417.

Song, Y., Lin, C. and Langston, C. (2005) A linkage measure framework for the real estate sector. *International Journal of Strategic Property Management*, 9, 121–143.

Song, Y., Liu, C. and Langston, C. (2006a) Extending construction linkage measures by the consideration of the impact of capital. *Construction Management and Economics*, 24, 1207–1216.

Song, Y., Liu, C. and Langston, C. (2006b) Linkage measures of the construction sector using the hypothetical extraction method. *Construction Management and Economics*, 24, 579–589.

Song, Y., Liu, C. and Li, H. (2007) Economic drive effect comparisons of the construction sectors between China and South Africa. *Proceedings of the Sixth Wuhan International Conference on E-Business*, Wuhan (China), 26–27 May, 2007.

Stahmer, C., Kuhn, M. and Braun, N. (1998) Physical input-output tables for Germany, 1990. Eurostat working paper n. 2/1998/B/1: Luxemburg.

Su, C., Lin, C. and Wang, M. (2003) Taiwanese construction sector in a growing maturity economy, 1964–1999. *Construction Management and Economics*, 21, 719–728.

Tilanus, C.B. and Rey, G. (1964) Input-output volume and value predictions for The Netherlands, 1948–1958. *International Economic Review*, 5, 34–45.

United Nations (1999) *Handbook of Input Output Tables Compilation and Analysis*. Studies in Method, Series F No. 74: New York.

Varian, H. (1992) *Microeconomic Analysis*, 3rd edition. Norton & Co: New York.

Weisz, H. and Duchin, F. (2006) Physical and monetary input-output analysis: what makes the difference? *Ecological Economics*, 57, 534–541.

Wolf, E.N. (1994) Productivity measurement within an input-output framework. *Regional Science and Urban Economics*, 26, 75–92.

Wu, X. and Zhang, Z. (2005) Input-output analysis of the Chinese construction sector. *Construction Management and Economics*, 23, 905–912.

Chapter 4

The scope of the construction sector

Determining its value

Les Ruddock and Steven Ruddock

Introduction

If the role of the construction sector in the overall economy of a country is to be fully understood, then it must be placed in perspective. Construction is an industry of major strategic importance and it is essential that the statistics produced for the industry are valid, reliable and comprehensive in their coverage. There has long been concern, though, that data on the construction industry are not as useful as for other sectors of the economy (Bon, 1990; Ruddock, 2000, 2003; Briscoe, 2006). A basic question must, therefore, be: How can construction statistics best be interpreted in order to capture the true scope of the industry?

In the Preface to the 'Pearce Report' (*The Social and Economic Value of Construction: The Construction Industry's Contribution to Sustainable Development*) (2003), the Chairman of nCRISP (the UK Construction Industry Research and Innovation Strategy Panel) stated that: 'The industry and its contribution to the UK economy and the health and well-being of UK society was neither fully understood nor adequately valued' (p ii). The United Nations defines construction as comprising 'economic activity directed to the creation, renovation, repair or extension of fixed assets in the form of buildings, land improvements of an engineering nature, and other such engineering constructions as roads, bridges, dams and so forth' (United Nations, 2001). Construction activity represents a significant share of the economies of most countries in terms of its contribution to GDP and total employment and is also an important market for materials and products produced by other sectors of the economy. Pearce considered construction both in its narrow sense (on-site construction activity) with its contribution to GDP at around 5 per cent and in a broader definition (including quarrying of construction raw materials, manufacture of building materials, sales of construction products and various associated professional services) making a contribution of about 10 per cent of GDP. The nCRISP Task Group considered that it could be taken even wider than that – to include land, property and facilities management but stopped, due to data availability and the fact that they 'had to stop somewhere' (Pearce, 2003: p. iii).

Construction statistics are usually of questionable quality, even in developed countries. Briscoe (2006) acknowledges the limitations of the traditional narrow measure of the construction industry in the UK and this view is echoed by other commentators concerned with the state of construction statistics in a wide range of countries. K'Akumu (2007), for instance, laments the fact that the narrowness of definition of construction statistics in a developing country such as Kenya makes the statistical portrait of the sector incomplete.

Construction's contribution to the economy

To consider construction activity to be merely the act of building is to take too narrow an interest. The productive issue to be solved by construction is more wide ranging and represents a considerable economic and social challenge. It is a question of producing and managing the living and working environment of the whole population. The entire built environment, as distinct from the natural environment, falls into the field of activity of construction.

On this basis, the principal aim is not merely to produce and manage structures for people's living and working environment, but rather '. . . to produce and manage the services rendered to end-users by these structures throughout their physical life cycle (production, use, improvement through to demolition)' (Carassus, 2004: p. 10). Valuing the construction industry, and assessing the importance of the industry to the economy as a whole is a difficult yet important task. Highlighting the importance of construction to the economy is a key point to ensuring it has a high priority on government agendas.

There is a distinction between the value *to* the economy and the value *in* the economy, both of which will vary vastly, dependent upon the definition of the sector used. The value *to* the economy considers the construction industry's use as a driver for growth, and as a catalyst for other industries to develop. A properly functioning construction sector can ensure a healthy economy, as the necessary infrastructure for economic activity to occur will be in place. The implicit value *in* the economy is more easily defined and assessed, and is measured as the contribution to national income. However, both assessments still rely on a consistent definition of the sector so that reliable and comparable estimates can be made.

The construction industry has an important role to play within the overall economy of any given country but how that role manifests itself will vary greatly from one nation to another. In developing countries, it is likely that the extraction of raw materials and on-site construction activity is most important, as the country seeks to set up a significant infrastructure, in the form of roads, railways and buildings. In developed countries, the emphasis is on professional services and the sale of end-products. It is also possible that a large repair and maintenance sector will emerge, the longer the main infrastructure has been in place, as potential customers seek to maintain and update current dwellings or work places rather than looking to new building altogether.

Broadly speaking, the construction industry is part of the process of creating and sustaining the built environment. In a narrow view, the construction industry is placed solely in the secondary sector, as this accounts for the transformation from materials into a final product. However, the reality is that the construction industry spans across the primary, secondary and tertiary sectors, as the process sees raw materials transformed into manufactured materials and then into a final product, with professional services and sale of products at the end of the line. The weightings of each part of the chain will vary from one country to another, skewed according to their level of development, with a higher concentration of primary and secondary sector firms in developing countries and more tertiary sector firms in developed countries.

As the construction sector increasingly diversifies, the need to define the industry becomes more apparent, with many traditional construction firms looking to broaden the scope of activities in which they participate, thus putting more emphasis on the need for a broader definition, with well defined boundaries. Any definition will need to encompass the entirety of the construction life cycle from the mining and extraction stage, through to the production of a final product and including the after sales service, all of which create part of the process of producing and maintaining the built environment.

The traditional perception of the contribution of the construction industry to the economy is based on the methodologies employed for the definition and measurement of construction activity according to international standards. Within this context, the limitations of the concepts used in this definition can be considered, and an analysis undertaken of the usefulness of the measures. Construction activity has changed in response to new demands over recent decades and an evaluation is made of an approach, focusing on construction activity to meet the changing needs of the economy and society. The role of built assets in the economic development of a nation needs to be considered and it may be that broader measures of the economic value of the built environment are needed in order to allow an assessment of the contribution of the built environment to quality of life and to enable the full value of the construction industry to be properly understood. A basic definition of the construction industry may not include other value-adding construction activities such as:

*   *Upstream* – manufacturing, mining and quarrying, architectural and technical consultancy, business services.
*   *Parallel* – architectural and technical consultancy.
*   *Downstream* – real estate activities.

Another factor that makes the construction sector of utmost importance to the economy is the role it plays in sustainable development. By the proper implementation of a sound infrastructure, a base for sustainable development can be laid in place. Coupled with ever advancing technologies in new build, and similar technologies applied to maintenance and alteration of existing builds, the construction industry plays a key role in ensuring a country can sustain a given level of development. Sustainability is increasingly becoming high priority to more and more countries. According to *Sustainable Development and the Future of Construction* (Bourdeau et al., 1998) the drive for sustainability identifies economic, social and cultural aspects as part of the sustainable construction framework, but special regard is given to ecological impacts on the environment. With more countries joining environmental pacts, and with fossil fuels and exhaustible materials for building becoming more scarce, the construction industry's part in sustainable development is becoming more important. Building projects which incorporate energy-saving schemes, such as advanced insulation, or natural energy-creating technologies such as solar panels, or indeed that use innovative materials in the physical build, contribute to sustaining the environment and thus aid the overall objective of sustainable development.

Diversification in construction

Not so long ago, construction firms dealt solely with the build of the actual project in hand. Over the years, however, these firms have begun to diversify both upstream and downstream. Many construction firms now run their own extractive operations, to provide themselves and others with the raw materials needed to create the manufactured materials that go into the on-site process. They have also moved towards providing services after the final product has been made. These include the sales of the products themselves and, in many cases, the security and maintenance of those products, including facilities management services.

The reasons for such diversification are varied. Some firms have moved into other areas, as it is cheaper for them to provide the service themselves than to pay for someone else to do it. Other firms have reacted to the volatility of the construction sector. By diversifying, they can ensure that their business will stay afloat. The construction industry has a high rate of firms becoming insolvent due to the high level of competition, especially within a regional context. Diversification means if one part of the business suffers due to the effects of high competition then another part can become the main focus of the organization. A diversified business may also be more attractive to potential clients, as a much fuller package can be offered from the initial design stage right through to the maintenance of the final product. Usually, where diversification within a firm exists, the firm can benefit from organizational economies. Indeed, the more the industry as a whole diversifies, the greater the potential benefit to be gained. In recent years, however, alternative firms have begun to offer similar services. Large companies, such as Marks and Spencer, now also offer services such as security, realizing they could provide the service for themselves rather than pay the firm who built their store or warehouse. Realizing the potential of such extra activities, these companies have now also started to offer the services outside in order to make themselves more diversified and gain from similar benefits as the construction firms have done.

This new development means the future for construction firms' diversification is unclear. They may continue to open up into alternative avenues, or, due to the higher competition they now encounter, they may withdraw and return to basic construction services again. The latter is less likely. The construction industry is a highly competitive environment and any advantage that can be gained over rival firms is likely to be taken. Competitiveness can be gained in varying ways. For existing firms, this competition may be based on price or quality. When so many firms are involved, an extra edge is often needed and this can be gained in two ways. Primarily, it is the diversification route, but some firms will rely on being at the forefront of technological advancements, so they can offer cutting edge designs, styles and materials to their clients. Both of these come under the umbrella of innovation; a firm that is innovative has a strong chance of survival in an extremely competitive environment. Either way the extent of this diversification needs to be considered in any evaluation of the sector.

Measuring the sector

Using Pearce's definition, the narrow sector consists solely of on-site assembly including repair work, which encompasses site preparation, construction of buildings and infrastructure, building installation and building completion. The broader definition consists of much more, including the supply chain for construction-related products, including the mining of construction materials and the manufacture of construction products. The broad definition also includes professional services such as management, architecture, design and facilities management. These two definitions are illustrated by Figure 4.1, which depicts the structure of the construction industry, and how the different components feed into the built environment.

Pearce states that: 'The wider definition has the virtue of drawing attention to the economic activities that directly depend on the narrower definition of the construction industry. The fortunes of these activities are critically interdependent with the fortunes of the contractors' (Pearce, 2003: p. 10). This highlights the ultimate importance of the wider view of construction, as all of the factors which make up the broader view are critical parts of the construction industry input into the built environment. The separate stages of the construction process, from mining and quarrying through to sales and management are all dependent upon one another.

In the UK, the narrow view yields an economic impact of around 5 per cent of GDP, and encompasses approximately 170,000 firms, whereas the broader view is roughly twice this amount contributing 10 per cent of GDP and including 350,000 firms (Pearce, 2003). There is a significant difference between the narrow and broad definitions and the resulting impact the construction industry has on the economy. However, both of these definitions are

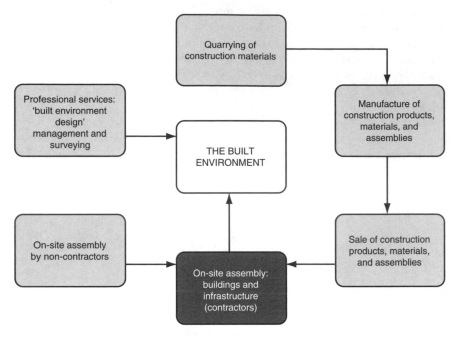

Figure 4.1 Broad and narrow industry definitions (The Pearce Report, 2003).

quite high level, including 'sectors' of the construction chain rather than individual tasks.

The mesoeconomic (or sector) approach

In addition to the Pearce Report, the case for a new approach to the valuation of construction activity has come from two other areas. Firstly, the International Council for Research and Innovation in Building and Construction (CIB) *Revaluing Construction* agenda focuses on improving the value of final construction output and requires that the totality of activities involved in the production of the built environment is reviewed. (See Ruddock (2007) for a holistic assessment of construction.) Secondly, Carassus (2004) proposes a framework (or mesoeconomic) system approach for understanding the construction sector. The rationale for this approach is based on the view that the role of the construction sector should be viewed in a wider context than that of the narrowly defined International Standard Industrial Classification (ISIC) definition of the industry. Figure 4.2 illustrates this approach and indicates the extent of the construction sector system.

In the decades following World War II, those countries developing a strong industrial base undertook large-scale infrastructure projects and housing schemes as a major part of their economic growth process. Towards the end of

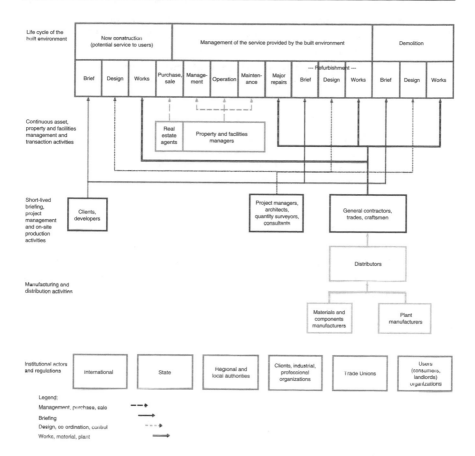

Figure 4.2 The construction sector framework.

Source: Carassus, 2004.

the twentieth century, however, approximately half of construction activity in these countries had become refurbishment and maintenance work. In this modern environment, quality of service has become of the essence. Outsourcing of property and facilities management services has reflected this concern, and expansion of built environment management over long periods has reflected this evolution.

The requirements of sustainable development focus on the need to deal with longer-term consequences, not only regarding the production of buildings but also the management of the building product over the whole life cycle. This focus on the service means that a new approach is needed to analysis of the function of the construction sector, with emphasis on the 'management of the service rendered by such works all along their life-cycle' (Carassus, 2004).

Economic analysis has to take into account such recent evolution and all the actors involved in the life cycle of building structures (not only procurement, design and production but also operation, maintenance, refurbishment and demolition). Construction sector analysis needs to go beyond just the construction firms to include the industry's professions and the materials industry as well as the service aspects, stock management organizations and the real estate sector.

Degree of 'influence' of the construction sector

In order to truly ascertain the full value of the construction industry and create a better defined description, a more detailed look at individual Standard Industrial Classification (SIC) – the internationally recognized listing of industries – codes is required. In Chapter 3, it was explained how input–output tables can be used in an examination of the inter-relationships between the productive sectors of an economy. At the heart of the input–output matrix is the intermediate demand section, which, in the case of the UK input–output accounting framework, displays transactions between 123 industry sectors. The columns of each sector consist of intermediate outputs. If a column element is divided by the column sum, technical coefficients are derived showing the amount spent on a particular input per £1 spent on total inputs.

The tables provide useful indicators for studying the input and output profiles of the construction sector. On the input side, they offer the possibility of tracing the overall interdependence of construction and its supplying industries, with the total input coefficients showing the effect of changes in the final demand for construction goods and services on any industry supplying the construction sector. Going back to the 1960s, there have been numerous studies analysing the input profile of the construction industry based on the direct requirements of the construction sector.[1]

Direct and total backward linkage indicators

Studies such as that of Pietroforte and Gregori (2003) show that the construction sector as a whole is characterized by relatively high backward linkage indicators, due to the fact that construction is largely an assembly industry. Direct backward linkage indicators represent the intermediate to total input ratios, i.e. the share of intermediate goods and services purchased by the sector per unit of their total purchases. When indirect purchases (either products incorporated into a product or products not incorporated into such a product but used by an industry in order to supply construction) are included, then total backward linkages (also termed output multipliers) can be calculated to show the effect of a one unit change in final demand of the construction sector on the generation of total output (both direct and indirect) in the national economy as a whole.

The construction sector generally has one of the highest output multipliers among all the sectors of a national economy. Using constant prices, in Pietro-forte and Gregori's aforementioned study of eight highly developed countries during the 1970s and 1980s, they calculated an overall average of 2.20 for construction. For the rapidly developing Chinese economy, Wu and Zhang (2000) estimated an output multiplier for the construction sector of 3.152 (making it the fifth highest ranking of the 17 sectors) in the whole economy in 2000. Interestingly, the latter also calculate an 'influence coefficient' for the construction sector (the ratio of the total input coefficient of the sector to the average total input coefficient of all sectors) and indicate that, with a value greater than 1 (at 1.262), its pull effect is greater than the average pull effect. High-output multipliers and backward linkage indicators for a sector have the potential to stimulate production in the many other sectors linked to it. The share of construction in direct employment generation may not be large, but its importance lies in its direct and indirect contribution through strong backward linkage effects. The fact that these 'pull effects' on the economy as a whole are amongst the highest can provide the key to government policies aimed at employment creation and maximising economic growth.

In the SIC, the narrow definition of the construction industry is indicated by SIC code 45. Using UK Input–Output Analysis (2005) and UK SIC 2003, SIC code 45 'Construction' gives a gross value added (GVA) as a percentage of total GVA of 6.2 per cent. If we look closely at the full SIC industry classification and the UK Input–Output tables we can establish which other SIC codes (or part SIC codes) fall within the boundaries of the broader definition. Table 4.1 indicates those areas including SIC 45 'Construction', which can be included in order to give a 'fuller view' of the scope of the construction sector, and to assist in giving a more appropriate valuation for the amount that the construction industry contributes to the economy. Based on the UK SIC (2003), the list of activities in Table 4.1 has been selected as those areas making significant contributions to 'construction' (with consideration of the views of the selected sources indicated).

For the UK, for 2003, an attempt to calculate construction sector GVA as a percentage of total GVA based on inclusion of the SIC components in Table 4.1 and making direct linkage calculations, yields a total GVA figure almost twice as large as that indicated in the broad 'Pearce definition'. (Values are calculated from the input–output tables to determine direct backward linkages and, therefore, only include the proportion of each category that is actually directly attributable to the construction sector). It is clear that, although the broad definition in the Pearce Report is useful, it still does not give the full picture. By only using high-level descriptions of 'sectors' within the construction industry, it is difficult to measure the total impact caused, as in some cases too much value may be attributed to the construction sector, but more critically many aspects are missed altogether. A more detailed breakdown of how the construction industry contributes to the economy also has the advantage of showing which areas are most important (value-added importance rather than actual importance within the chain) to the economy.

Table 4.1 Principle activities related to construction activity (Based on UK SIC 2003; derived from: Ive and Gruneberg, 2000 (Part 1, Appendix); Pearce, 2003 (Statistical Appendix. Table 16); Department of Business, Enterprise and Regulatory Reform: Definitions of Construction Sector Services and All Property)

Section		Group/class and description	
F	Construction	45	Construction
C	Mining and quarrying	14.1	Quarrying of stone
		14.2	Quarrying of sand and clay
D	Manufacturing	17.51	Manufacture of carpets and rugs
		20.1	Saw milling etc. of wood
		20.2	Manufacture of plywood and board
		20.3	Manufacture of builders' carpentry and joinery
		24.3	Manufacture of paints, varnishes etc.
		25.23	Manufacture of builders' ware of plastic
		26.1	Manufacture of glass and glass products
		26.22	Manufacture of ceramic sanitary fixtures
		26.3	Manufacture of ceramic tiles and flags
		26.4	Manufacture of bricks, tiles and construction products in baked clay
		26.5	Manufacture of cement, lime and plaster
		26.6	Manufacture of articles of concrete, plaster and cement
		26.7	Cutting etc. of stone
		28.1	Manufacture of metal structures etc.
		28.22	Manufacture of central heating radiators and boilers
		28.63	Manufacture of locks and hinges
		29.23	Manufacture of non-domestic cooling and ventilation equipment
		29.52	Manufacture of machinery for mining, quarrying and construction
		31.3	Manufacture of insulated wire and cables
		31.5	Manufacture of lighting equipment and electrical lamps
G	Trade	51.53	Wholesale of wood, construction materials and sanitary equipment
		51.54	Wholesale of hardware, plumbing and heating equipment and supplies
		51.82	Wholesale of mining, construction and civil engineering machinery
J	Financial intermediation	65.12	Banks and building societies
		65.21	Financial leasing
		65.22/3	Activities of mortgage finance companies
		65.23/5	Activities of venture and development capital companies
		66.01	Life insurance
		66.02	Pension funding

K	Real estate, renting and business activities	70.11/12	Development and selling of real estate/buying and selling of own real estate
		70.20	Letting of own property
		70.31/32	Real estate agencies/management of real estate on a fee or contract basis
		71.32	Renting of construction and civil engineering machinery and equipment
		74.20. 1–2	Architectural activities/urban planning and landscape architectural activities
		74.70	Industrial cleaning

The main contributors to the economy of the UK construction sector are the construction work itself, various construction products, architectural activities and real estate activity. These components make up 95.2 per cent of the overall construction contribution to GVA. This type of knowledge can aid government aims and objectives, and could, for instance, influence a shift in the concentration of regulations, in order to ensure that the stronger sectors of the industry thrive, but are also run legitimately with safety, health and environmental issues also considered. This can help a government to plan ahead, and to be prepared for the changes that will occur and to incorporate performance indicators that will take into account the changing nature and the full expanse of the industry.

A similar analysis can be applied to the construction sectors of other economies. Based on ISIC rev.3, Figure 4.3 shows an estimated measurement of value added by the construction sector, in 20 European countries. Due to differences in accounting practices, classification and aggregation of industry sectors, such international comparisons are fraught with difficulties. Whilst the results are purely indicative and approximate, they nevertheless give provide a fairly consistent picture of a construction sector, which has a considerable sphere of influence in the overall economy.

The contribution of the construction industry to economic development

The view that, in the early stages of economic development, the share of construction increases but ultimately declines (in relative terms) in industrially advanced countries was put forward by Bon (1992) and has been empirically tested by, amongst others, Ruddock (2000) and Ruddock and Lopes (2006). An explanation of this inverted 'U' relationship between construction activity and the level of income per capita is succinctly given by Tan (2002: p. 595): 'In low income countries, construction output is low. As industrialisation proceeds, factories, offices, infrastructure and houses are required, and construction as a percentage of gross domestic product reaches a peak in middle income

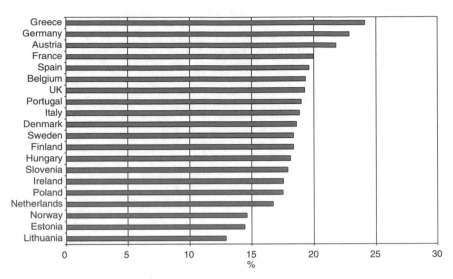

Figure 4.3 Size of the construction sector in twenty European countries (as percentage of GCA) based on homogenized ISIC classifications. All years 2000 except Slovenia 2001; Norway 2002. (Data sources: National Statistics databases.)

countries. It then tapers off as the infrastructure becomes more developed and housing shortages are less severe or are eliminated.' An empirical Construction Industry Comparative Analysis study of nine developed countries undertaken by the CIB project group (Carassus et al., 2006) showed that, after allowing for cyclical fluctuations, the general trend in construction activity in very developed countries is for construction activity to be in relative decline.

If the availability of adequate, reliable data permitted an assessment of the broader construction sector and its linkages across several countries at different stages of development, this would help to better explain the link between economic development and the construction industry, and could support the premise that developing countries rely on differing aspects of the chain that makes up the construction industry framework. This should involve comparisons of the proportions of the ISIC codes attributable to the construction industry, to check the degree of consistency across countries regardless of stage of development. This type of information would be of great value to economic planners in countries, where detailed data were limited or unavailable, if there were some validity to the premise that the relationships followed a similar pattern.

Times series analysis using such data would give a good indication of where the industry may go in the future. A relatively slow-changing industry for a prolonged period is less likely to suddenly change than one that has been changing constantly over that time. A second benefit would be to the developing countries that are at a different stage of the construction life cycle, and so can look at how a developed country's construction industry evolved over time, and make provisions for similar changes (taking into account the current economic climates) in their own industry.

Taking an example of the application of the construction sector approach in the context of a developing country; in Sri Lanka, it was not surprising that, post-tsunami, the 'official' construction sector growth rate outstripped the growth in GDP (see Figure 4.4). In the supporting sectors growth was particularly strong.

The notion of attempting to measure the value of the 'ancillary' parts of the construction sector in a developing country can be much more difficult due to a lack of comprehensive information, particularly inter-industry (input-output) data. Using a methodology, in which the inter-sectoral relationships derived from the aforementioned study of developed countries are used to determine coefficients gave an estimate of 16.8 per cent for the construction sector's share of GVA in the first quarter of 2007 (Ruddock, 2008). This measurement is based upon a snapshot, single-year assessment and a consistent set of time-series data would, of course, be needed to give the representation more validity. However, it is interesting to note that this percentage figure fits within the range of values observed in the developed countries shown in Figure 4.3.

Conclusions

Defining the scope of the construction sector is a key component for valuing the industry, for measuring the sector's importance to the economy, for understanding construction's role in sustainability and it is essential for making provisions for the future.

The make-up of the industry itself is more complex than it may first seem, a simple 'narrow' definition does not suffice, as a whole chain of stages is required to feed into the end-product – *the built environment*. This chain needs to not only be considered at a high level, but also at a more detailed level in order to pick out the specific areas that are attributable to the construction sector. Using these specific aspects, a value for the amount the construction industry contributes to the economy in terms of GVA can be calculated, which will be a far greater figure than the usual estimations of 5–6 per cent for the narrow definition and 10 per cent for the broader one. However, the construction industry contributes much more than that. It helps improve growth and aids sustainability through improved infrastructure.

The proportions of what make up the industry are likely to be different for countries at different stages of development. Developed countries are more heavily dependent upon the later stages of the construction life cycle, whereas developing countries are much more likely to rely upon the earlier stages of the production of a built environment and the production of the build itself, especially large-scale infrastructure build. These inherent differences lead to differing government objectives, drivers and regulations, in order to gain the maximum benefit from the industry, and to ensure that its activities are carried out in an appropriate manner, whilst planning and building for the future.

The construction industry is a complex one, with many interactive levels, and several 'stages of production' that create an end-product. Each stage is also made up of various aspects. It is when this detailed level of the stages is looked at, that it is possible to ascertain the full make-up of the industry and thus give it a full and comprehensive definition, that can be used to value the industry as a single unit, and highlight the impact the construction sector has on the economy as a whole.

Note

1   Chapter 3 has more detail on the development of such studies.

References

Bon, R. (1990) The World Building Market 1970–85. *Building Economics and Construction Management: Proceedings of the CIB W65 Symposium*, Sydney.

Bon, R. (1992) The future of international construction: secular patterns of growth and decline. *Habitat International*, 16(3), 119–128.

Bourdeau, L., Huovila, P., Lanting, R. and Gilham, A. (1998) *Sustainable Development and the Future of Construction*. CIB Report No.25: Paris.

Briscoe, G. (2006) How useful and reliable are construction statistics? *Building Research and Information*, 34(3), 220–229.

Carassus, J. (ed) (2004) *The Construction Sector System Approach: An International Framework*, Report by CIB W055–W065 'Construction Industry Comparative Analysis' Project Group. CIB: Rotterdam.

Carassus, J., Andersson, N., Kaklauskas, A., Lopes, G., Manseau, A., Ruddock, L. and de Valence, G. (2006) Moving from production to services: a built environment cluster framework. *International Journal of Strategic Property Management*, 10(3), 169–184.

European Commission, Eurostat. *Input-Output Tables*.

Ive, G.J. and Gruneberg, S.L. (2000) *The Economics of the Modern Construction Sector*. Macmillan Press: London.

K'Akumu, O.A. (2007) Construction statistics review for Kenya. *Construction Management and Economics*, 25, 315–326.

Pearce, D. (2003) *The Social and Economic Value of Construction: The Construction Industry's Contribution to Sustainable Development*. nCRISP: London.

Pietroforte, R. and Gregori, T. (2003) An input-output analysis of the construction sector in highly developed economies. *Construction Management and Economics*, 21, 319–327.

Ruddock, L. (2000) *An International Survey of Macroeconomic and Market Information on the Construction Sector: Issues of Availability and Reliability*. RICS Research Paper Series, Volume 3, Number 11. RICS: London.

Ruddock, L. (2003) Measuring the global construction industry: improving the quality of data. *Construction Management and Economics*, 20, 553–556.

Ruddock, L. (2007) A holistic view of the value of construction. In: Barrett, P. (ed) *Revaluing Construction*. Blackwell: London.

Ruddock, L. (2008) The Importance of the Construction Sector: Measuring its Value. *Proceedings of CIB W89 International Conference on Building Education and Research*, Kandalama, Sri Lanka.

Ruddock, L. and Lopes, J. (2006) The construction sector and economic development: the 'Bon' Curve. *Construction Management and Economics*, **24**, 717–723.

Tan, W. (2002) Construction and economic development in selected LDCs: past, present and future. *Construction Management and Economics*, **20**(7), 593–599.

UK Input-Output Analysis (2005) The Stationery Office: London.

UK SIC (Standard Industrial Classification of Economic Activities) (2003) The Stationery Office: London.

United Nations (2001) *International Recommendations for Construction Statistics*, Series M, No.47, Rev.1. United Nations: New York.

Wu, X. and Zhang, Z. (2005) Input-output analysis of the Chinese construction sector. *Construction Management and Economics*, **23**, 905–912.

Chapter 5

Investment in construction and economic growth

A long-term perspective

Jorge Lopes

Introduction

The role of construction in economic growth and development has been addressed by various writers and international bodies, many of whom have focused on developing countries (Turin, 1973; World Bank, 1984; Wells, 1986; Bon, 1992). Bon (1992) analysed the changing role of the construction sector at various stages of economic development and presented a development pattern for the industry based on the stage of development of a country's economy. The main aspects of the proposition were that, in the early stages of the economic development, the share of construction in gross domestic product (GDP) increases but ultimately decreases in industrially advanced countries. Turin and Wells, using cross-country comparisons, both found an association between construction investment and economic growth. That finding was consistent with the classical approach in growth theory in which physical capital formation is the main engine of economic growth and development. In the aftermath of the 1979–1980 oil shock and the international financial crisis that followed in 1981, most of Sub-Saharan African countries experienced, until the mid-1990s, a decreasing growth in per capita national income, despite heavy investment in construction and other physical capital over the preceding decade.

Another approach to the theory of growth (Romer, 1990; Barro, 1991; Olsen, 1996) was emerging. Following this endogenous growth theory, endogenous policy changes (e.g. macroeconomic stability, investment in human capital, research and innovation) play an increasing role in the development process. De Long and Summers (1991), using data from the *United Nations Comparison Project* drawn from 61 countries representing all stages of economic development, found that machinery and equipment investment have a strong association with economic growth. Further, they put forward evidence that structures investment is only weakly associated with growth. World Bank (1994) posited that, rather than the quantity of infrastructures, the main concern in developing countries should be the improvement of the quality of infrastructures. Thus, it is reasonable to argue that this would be achieved through an adequate maintenance of existing infrastructure stocks and by prioritizing investments that modernize production and enhance international competitiveness. In the *Structural Adjustment Programme* for Africa, The World Bank and its affiliates seemed also to follow the view that investment should accompany economic growth.

The next part of this chapter presents the statistical sources and the indicators of the economic activity chosen for the analysis. This is followed by a historical review of the process of economic growth. The chapter then presents and analyses the pattern of development of construction investment in the world economy as well as in the world regional groups. A concluding remark finalizes the analysis.

Statistical sources and methodology of data collection and presentation

The main statistical sources used in this analysis are the most recent edition of the *Yearbook of National Account Statistics: Main Aggregates and Detailed Tables* from the United Nations and *World Development Report* from the World Bank (1992). The Internet site of the UN Statistical Office presents data on GDP and its components both in the expenditure and production approaches. This publication presents various sets of economic series detailing the evolution of GDP and its components in different statistical formats over the long period 1970–2006, both in the world, world regions and countries: at current prices in national currencies; constant 1990 prices in national currencies; current prices in US dollars; constant 1990 prices in US dollars. The indicators of economic activity analysed are: GDP; construction value added (CVA); and gross fixed capital formation. Unfortunately, data on gross fixed capital formation in construction (GFCFC) are not provided in the UN publication. Thus, CVA is used as a proxy for analysing the evolution pattern of construction investment across the world and world regions. Indeed, as pointed out in Lopes (1997), since World War II, when international bodies, particularly the United Nations, started publishing data on the construction sector, there has been a remarkable uniformity across countries on the value of 50 per cent as the average contribution of capital formation in construction to a country's domestic investment. As construction value added is roughly a half of the former, it appears reasonable that CVA can be used as a surrogate measure of construction investment. In order to facilitate international comparison as well as for aggregation purposes, constant 1990 prices in US dollars are used. With respect to the investigation of the relationship between the construction sector and economic development according to a country's (or group of countries') stage of economic development, gross national product (GNP) per capita for the bench mark year 1990 has been chosen. These data are provided by the World Bank (1992). *The World Development Report 1992* presents the following definitions:

- *Income group*: the economies are divided according to 1990 GNP per capita. The groups are: low income economies (LIEs), US$610 or less; lower-middle income (LMIEs), US$611–US$2,465; upper-middle income (UMIEs), US$2,412–US$7,619; and high income economies (HIEs), US$7,620 or more.
- *Subgroup*: LIEs are further divided by size, and HIE by membership of OECD.
- *Region*: economies are divided in five major regions and eight additional subregions.

A historical review of the processes of economic growth

This section presents a review of the historical experience of the economic growth of nations, especially of today's advanced industrial countries, in order to have a full picture of the nature of the economic factors that are conducive to long-term economic growth and development. Data presented here generally cover the period between 1870 and 1984, and in some cases go back as far as 1750. Growth accountancy owes much to the work of Simon Kuznets, who received the Nobel Prize in Economics in 1971 for his pioneering work in the measurement and analyses of the historical process of economic growth in developed countries (Todaro, 1992). Most modern studies on growth accountancy (Denison, 1985; Hickman and Coen, 1987; Maddison, 1987; Chenery et al., 1986, to name but a few) and the modern methodology of national accounts are based on the conceptual framework prepared by Kuznets. Another great student of growth accounting is Angus Maddison. In his monumental work (*The World Economy: A Millennial Perspective*), Maddison (2006) provides a fascinating accounting exercise on, in Lucas' (1988) words, 'the mechanics of the economic growth of nations from 0 AD until the present time'.

Trends in the growth of national product

The national income or total net product of a country is the sum of all goods and services during a given period, usually 1 year, adjusted for duplications, and net for any commodity consumed in the process of production.

Estimates made by Kuznets (1968) on a sample of countries representing 80 per cent of the world population in 1950 found that over 30 per cent had per capita income of less than $50 and almost one quarter, between $50 and $100 (US dollars, 1949 constant prices). On the top of the income per capita pyramid were the US, the UK, the Scandinavian countries, the Netherlands, France, Canada, Australia and New Zealand, which all had a per capita income of $600 or more. Taking in account the fact that the countries in which data were not available were former colonies, it is reasonable to assume that more than 60 per cent of the world population had a per capita income of less than $100 in 1950 (Kuznets, 1968).

As in 1984, and using the modern division of the world economy (*World Development Report* – World Bank, 1986), the average per capita income (US dollars, 1984 current prices) range varied from $260 in low income economies, to $1,250 for middle income economies and around $2,000 for the former East European (excluding Yugoslavia) non-market economies. Again, on top of the per capita income pile were the high income economies of the OECD with an average income per capita of $11,430. Portugal, Greece and Turkey, which belong to OECD, are included in the upper-middle income economies. As regards the long-run evolution of the national output, data are only available for the most developed countries. Tables 5.1 and 5.2 present, respectively the evolution of GDP and GDP per capita (in international dollars) of four Western European countries (France, Germany, the Netherlands and the UK), Japan and US, over the long-term period 1870–1984. Data are taken from Maddison (1987). It can be seen that the rate of growth throughout the period of analysis was increasing. Particularly striking is the growth process in 1950–1973. During this period, GDP per capita in the US had an increase of more than 50 per cent and in the UK almost 100 per cent. Japan increased its per capita GDP by more than fourfold and the remaining countries more than doubled their 1950 level of GDP per head. The process of convergence by the five countries to the leader country (US) was also accelerating. However, Table 5.1 also shows that the rate of growth

Table 5.1 GDP Levels (1984 International $) in selected industrial countries: 1870–1984 (Source: Maddison, 1987)

Year	France	Germany	Japan	Netherlands	UK	US
	($ billion)					
1870	59.27	33.98	19.28	8.26	77.95	78.61
1913	119.99	111.75	54.76	20.33	174.78	454.53
1950	173.49	179.22	124.34	49.40	281.04	1,257.86
1960	271.03	387.21	295.178	76.99	372.80	1,735.86
1973	547.98	675.49	976.50	142.20	556.60	2,911.78
1984	694.70	811.6	1,468.40	168.90	625.20	3,746.50

Table 5.2 GDP per capita (1984 International $): 1870–1984 (Source: Maddison, 1987)

Year	France	Germany	Japan	Netherlands	UK	US
	($)					
1870	1,542	1,336	560	2,290	2,671	1,962
1913	2,878	2,737	1,060	3,298	4,101	4,657
1950	4,147	3,600	1,486	4,884	5,000	8,261
1960	5,933	6,985	3,136	6,703	7,093	9,608
1973	10,514	10,899	8,987	10,581	9,902	13,741
1984	12,643	13,235	12,235	11,710	11,068	15,829

in all countries had started to decelerate since 1973. Average GDP growth for the four European countries and Japan fell from 5.6 per cent p.a. in 1950–1973 to 2.1 per cent p.a. thereafter, and for the U.S. from 3.7 per cent to 2.3 per cent.

For the less developed countries, there is no consistent set of data available on the national output prior to 1950. Reynolds (1985) found that 23 out of 41 countries analysed reached the *turning point* at various points in time between 1850 and 1950, most of them between roughly 1890 and 1914, which corresponded to a period of worldwide economic boom. These successful countries experienced, during this period, a sustained rise in their per capita income, and in some of them (e.g. Argentina before 1914), the development pattern was similar to that of the HIEs during the same period. It should be noted that data used by Reynolds (1985) concern agricultural production, exports and population. Another relevant characteristic of the development process pertaining to the LIEs is that some of these countries (e.g. Nigeria, Zambia and Zimbabwe) reached the turning point in their colonial status.

From the period 1950 (when the United Nations and the World Bank started to publish national accounts statistics of their member sates) onwards, there is a reasonable amount of data on the national output and its components pertaining to the LIEs. In the post World War II period, the rate of growth of GNP per capita in LIEs has been roughly the same as in HIEs. However, significant differences in the growth process within the LIEs can be observed. In the period 1965–1990, the average annual growth rate of the GNP per capita was 0.2 per cent in Sub-Saharan Africa compared to 5.3 per cent in East Asia & Pacific. In the same period, the average annual growth rate was 2.5 per cent and 2.4 per cent for, respectively, low and middle income economies and OECD members (*World Development Report* – World Bank, 1992).

Trends in capital formation

As stated before, capital formation as one component of the national product is measured by net or gross additions to the stock of construction and of production's equipment and net additions to household inventories. Again, the long-term data concern the present-day more developed countries.

Table 5.3 presents the evolution of the gross fixed capital formation in six advanced industrial countries in the period 1913–1984. It is observed that, in line with the growth in GDP and GDP per capita (see Tables 5.1 and 5.2.), gross fixed capital formation increased in all countries during the period of analysis. It is also shown that the country which experienced the fastest rate of economic growth – measured either in GDP or GDP per capita – during the period of analysis (Japan) was the same in which the growth rate in capital formation was highest. However, as suggested in Table 5.3, the rate of growth in the latter indicator in all countries started to decelerate since 1973, following the pattern of the GDP.

The association between capital formation and the evolution of GDP is

Table 5.3 Total gross fixed capital stock (asset weights) at midyear: 1913–1984 (Source: Maddison, 1987)

	France	Germany	Japan	Netherlands	UK	US
Year	1950 = 100					
1913	–	–	56.66	45.99	65.17	51.34
1950	100.00	100.00	100.00	100.00	100.00	100.00
1960	126.62	163.09	156.06	138.46	128.51	137.18
1973	226.56	331.78	584.77	238.85	210.13	214.74
1984	348.67	478.89	1,225.73	344.53	276.06	290.66

Note: – not available.

Table 5.4 Capital productivity growth: 1913–1984 (Source: Maddison, 1987)

	1913–1950	1950–1973	1973–1984	1913–1984
	(Annual average compound growth rates)			
France	0.12	1.50	−1.82	0.23
Germany	0.56	0.57	−1.71	0.20
Japan	0.69	1.39	−3.41	0.28
Netherlands	0.31	0.85	−1.83	0.15
UK	0.13	−0.26	−1.45	−0.24
US	0.96	0.34	−0.47	0.55
Average	0.46	0.73	−1.78	0.20

better understood looking at the Table 5.4. As pointed out by Maddison (1987: p. 656), the close correlation between capital and output movements over the long run is the reason simple regressions find physical capital such a powerful explanation of growth.

As indicated in Table 5.4, the average of capital productivity (the difference between the compound rate of increase in output and the rate of increase in capital input) of the six countries was 0.20 per cent p.a. over the period 1913–1984. Further, capital productivity was positive in the period referred to for all countries, except for the UK.

However, contrary to labour productivity which seldom declines on a short-run basis, and in the long run is always positive, capital productivity is moderately positive and can be significantly negative in depressed economic conditions (ibid., p. 657). In the period 1973–1984, capital productivity was negative in all six countries and their average was −1.78 per cent p.a. This suggests that capital formation is not the only determinant of growth. It is also shown that the US had, among its counterparts, the highest capital productivity in the long period 1913–1984 and has been the leader country since 1913. It is not surprising that data based on the growth experience of the US was the basic material used in Solow's (1956) neoclassical model of economic growth.

The relationship between investment in construction and economic growth

Data

As referred to earlier, the indicator used as a proxy for construction investment is construction value added. CVA is calculated the same way as in any other sector, but includes only the activities of the construction activity proper. For example, it excludes the building materials industry, which is accounted in the manufacturing sector. The main indicator of economic activity used in this study is GNP per capita. It adjusts the growth in the economy with the growth in population. It is a better indicator of a country's welfare, particularly in developing nations, where the growth rate of population has been, since World War II, roughly twice as high as in developed economies.

Using data adapted from the UN Yearbook of National Accounts Statistics (United Nations, 2008 internet edition) and World Bank (1992), data are presented for gross domestic product (GDP), gross fixed capital formation (GFCF) and construction value added (CVA) at constant 1990 US$ (Figure 5.1). GNP per capita is presented for the benchmark year 1990.

Cross-matching sources, data is available for 93 countries and these countries can be split into three groups according to the level of GNP per capita in 1990. Tables 5.5–5.7 illustrate these three groups: Group I – low income countries (LICs); Group II – middle income countries (MICs); Group III – high income countries (HICs).

The division according to the level of GNP per capita in 1990 is used as a proxy for the level of economic development for these countries. This categorization is not an attempt to apply labels of least developed through to most

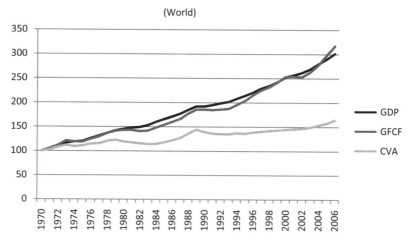

Figure 5.1 Volume indices of GDP, GFCF, CVA at constant 1990 US$ (1970 =100).

Table 5.5 GNP per capita and share of CVA in GDP for selected years (Group I)

Region	Country	GNP per capita cur. US$ (1990)	CVA/GDP (%) (1990)	CVA/GDP (%) (2006)
Sub-Saharan Africa	Benin	360	3.19	3.91
	Burkina Faso	330	4.66	5.49
	Burundi	210	3.35	3.11
	Central African Rep.	390	2.81	2.97
	Chad	190	1.69	1.31
	Ghana	390	3.30	3.56
	Kenya	370	2.93	2.53
	Lesotho	530	14.68	12.44
	Madagascar	230	1.19	2.83
	Malawi	200	4.96	5.10
	Mali	270	2.91	4.92
	Mauritania	500	4.78	8.95
	Mozambique	80	5.19	7.33
	Niger	310	2.45	2.55
	Nigeria	290	1.69	2.40
	Rwanda	310	6.78	9.18
	Somalia	120	3.63	4.89
	Sierra Leone	240	2.11	3.96
	Sudan	190	6.03	5.00
	Tanzania	110	4.76	7.81
	Uganda	220	4.84	7.58
	Zaire	220	5.00	4.24
	Zambia	420	2.58	3.40
East Asia and Pacific	China	365	4.44	4.39
	Cambodia	181	2.39	5.62
	Indonesia	570	5.09	5.42
	Lao PDR	200	2.92	2.61
	Vietnam	380	3.84	5.55
South Asia	Bangladesh	210	5.65	8.59
	India	350	4.96	5.24
	Nepal	170	8.65	9.13
	Pakistan	380	2.76	2.08
	Sri Lanka	470	4.71	5.27

developed to the three groups but represents an arbitrary proxy for the classification based simply on GNP per capita (Ruddock and Lopes, 2006).

In order to give a general picture of different regions of the world, these three groups were divided into eight additional subgroups according to the World Bank nomenclature of world subregions. Figures 5.2–5.11 illustrate the subgroups. Strictly speaking SSA-LICs comprises the countries south of the Saharan desert except for a few countries that belonged to the middle income range in 1990. East Asia and Pacific LICs comprises countries of the low income range

Table 5.6 GNP and share of CVA in GDP for selected years (Group II)

Region	Country	GNP per capita cur. US$ (1990)	CVA/GDP (%) (1990)	CVA/GDP (%) (2006)
Sub-Saharan Africa	Botswana	2,040	7.28	5.14
	Cape Verde	890	11.91	8.76
	Congo, Rep	1,010	2.99	5.01
	Cote d'Ivoire	750	1.79	3.29
	Gabon	3,330	6.71	6.53
	Mauritius	2,250	5.62	5.96
	South Africa	2,530	2.98	2.62
East Asia and Pacific	Korea, Rep	5,400	10.18	6.54
	Malaysia	2,320	3.90	2.82
	Philippines	730	6.02	4.32
	Thailand	1,420	5.95	2.42
Eastern Europe and Central Asia	Albania	1,290	6.63	6.47
	Bulgaria	2,250	6.96	4.29
	Hungary	2,780	5.55	5.33
	Poland	1,690	8.79	11.81
	Romania	1,640	5.41	7.60
	Turkey	1,630	6.30	4.61
North Africa & Middle East	Algeria	2,060	10.31	10.83
	Iran	2,490	5.30	5.92
	Morocco	950	4.52	4.39
	Saudi Arabia	7,050	6.53	6.71
	Syria	1000	3.77	4.00
	Tunisia	1440	4.06	4.81
South America, Central America & Caribbean	Argentina	2,370	4.44	3.62
	Bolivia	630	3.07	2.35
	Brazil	2,680	7.13	5.62
	Chile	1,940	7.36	6.47
	Colombia	1,260	4.74	3.86
	Costa Rica	1,900	4.27	3.97
	Dominican Republic	830	6.66	7.27
	Ecuador	980	3.86	3.49
	El Salvador	1,110	3.47	3.61
	Guatemala	900	4.40	4.17
	Jamaica	1,500	11.70	8.84
	Mexico	2,490	3.59	3.48
	Panama	1,830	1.91	4.02
	Paraguay	1,110	5.45	5.49
	Peru	1,160	3.67	4.42
	Trinidad & Tobago	3,610	9.21	10.16
	Uruguay	2,560	3.36	3.10
	Venezuela	2,570	5.72	4.90

Table 5.7 GNP and share of CVA in GDP for selected years (Group III)

Region	Country	GNP per capita cur. US$ (1990)	CVA/GDP (%) (1990)	CVA/GDP (%) (2006)
HICS OECD East Asia & Pacific	Australia	17,000	6.20	6.96
	Japan	25,430	9.87	5.73
	New Zealand	12,580	3.92	4.48
HICs OECD Europe	Austria	19,060	6.14	6.42
	Belgium	15,514	5.01	4.59
	France	19,490	5.12	3.68
	Denmark	22,080	4.41	3,67
	Finland	26,040	7.19	4.16
	Germany	22320	5.51	3.26
	Ireland	9,550	4.87	8.10
	Italy	16,830	5.62	5.15
	Netherlands	17,200	5.21	3.81
	Norway	23,120	4.29	3.04
	Spain	11,020	8.03	8.16
	Sweden	23,660	5.90	4.01
	Switzerland	32,680	8.18	6.14
	UK	16,100	6.22	5.03
OECD Americas	Canada	20,470	6.30	5.34
	USA	21,790	4.48	3.42

including China and Indonesia. East Asia and Pacific MICs comprises middle income countries of East Asia such as the republic of Korea, Philippines and Malaysia. OECD East Asia and Pacific comprises Australia, New Zealand and Japan. South Asia is composed of the low income countries of that region including India, Pakistan and Sri Lanka. Eastern Europe and Central Asia is composed of the middle income countries of Eastern Europe, apart from the countries that were formed after the breakup of the former Soviet Union. OECD Europe comprises 14 high income European countries that are members of the OECD. Middle East and North Africa comprises middle income countries of that region for which data are available. The Rest of the Americas comprises countries of South America and the Caribbean, both in the middle income range. OECD America is composed of United States and Canada.

Analysis

Table 5.8 shows the evolution of GDP, GFCF and CVA in the world for the period 1970–2006. The value of the construction output (measured as value added) reached US$1,578 billion (at 1990 constant prices) in 2006 – more than the national output of the UK. Assuming that CVA is roughly a half of capital

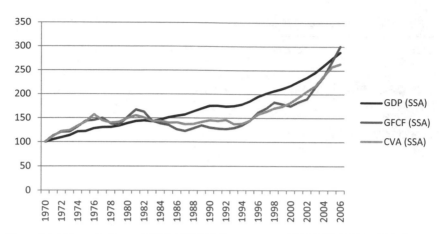

Figure 5.2 Volume indices of GDP, GFCF, CVA at constant 1990 US$ (1970 =100) (SSA).

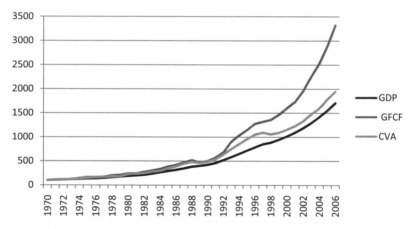

Figure 5.3 Volume indices of GDP, GFCF, CVA at constant 1990 US$ (1970 =100) (East Asia and Pacific LICs).

formation in construction, total construction investment in the world had a total value of about US$3.2 trillion (at 1990 constant prices) in 2006. It also can be seen that both indicators increase throughout the period but with different patterns: GDP and GFCF increased by more than 200 per cent in the period 1970–2006 whereas CVA increased by just over 60 per cent in the same period (as illustrated in Figure 5.1). Now looking at the picture at regional levels, it can be seen that the rate of growth of the CVA was higher in the less developed countries, particularly in East Asia and Pacific LICs where CVA measured at 1990 constant prices rose more than nineteenfold. Notice also the spectacular

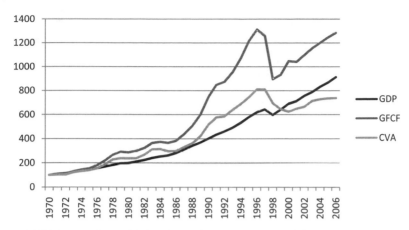

Figure 5.4 Volume indices of GDP, GFCF, CVA at constant 1990 US$ (1970 =100) (East Asia and Pacific MICs).

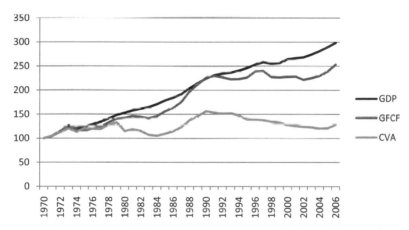

Figure 5.5 Volume indices of GDP, GFCF, CVA at constant 1990 US$ (1970 =100) (OECD East Asia and Pacific).

growth in GFCF – particularly in North America and the low income countries of East Asia and Pacific and North America.

Taking into account temporal fluctuations, this behaviour, both at group and world levels, tends to corroborate the observations made by Turin (1973), that the share of construction value added is generally between 3 and 8 per cent of GDP, an important pattern of the construction industry activity that has remained from the 1960s onwards.

Looking at countries grouped according to their economic development status, it can be seen in Tables 5.9–5.11 that the development pattern of CVA is

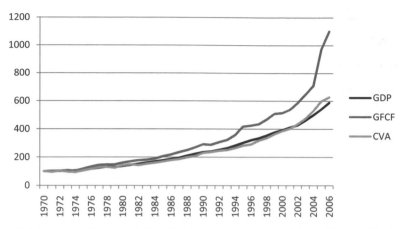

Figure 5.6 Volume indices of GDP, GFCF, CVA at constant 1990 US$ (1970 =100) (South Asia).

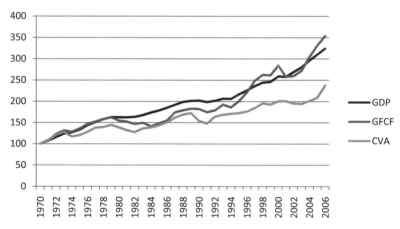

Figure 5.7 Volume indices of GDP, GFCF, CVA at constant 1990 US$ (1970 =100) (Eastern Europe and Central Asia).

markedly differentiated according to the level of economic development. The share of CVA in GDP in Group I increased over the whole period from 3.86 per cent in 1970 to 5.19 per cent in 2006. In Group II, the same indicator of the construction industry increased slightly in the period 1970–1980, decreased in the period 1980 per cent2000, and remained practically stagnant from 2000 onwards. In the high income countries, the construction value added decreased relatively but not absolutely. This fact does not corroborate Bon's (1992) assumptions that in the early stages of the economic development, the share of

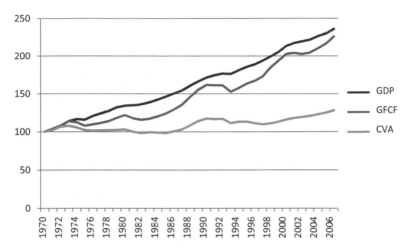

Figure 5.8 Volume indices of GDP, GFCF, CVA at constant 1990 US$ (1970 =100) (OECD Europe).

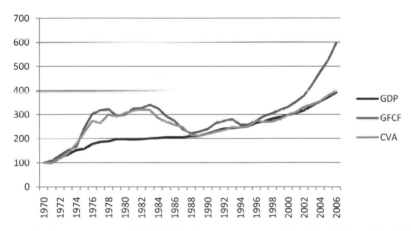

Figure 5.9 Volume indices of GDP, GFCF, CVA at constant 1990 US$ (1970 =100) (Middle East and North Africa).

construction in GDP increases but ultimately decreases (not only relatively but also absolutely in industrially advanced countries).

The pattern experienced by the MICs are worthy of note. It is shown in Table 5.10 that from 1990 onwards, the share of construction in GDP remained practically stagnant, at around 5.5 per cent of GDP. This suggests that construction does not lead economic growth but rather should follow the growth pattern of the general economy.

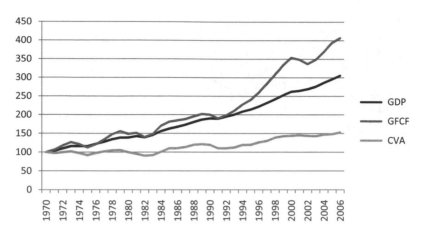

Figure 5.10 Volume indices of GDP, GFCF, CVA at constant 1990 US$ (1970 = 100) (OECD Americas).

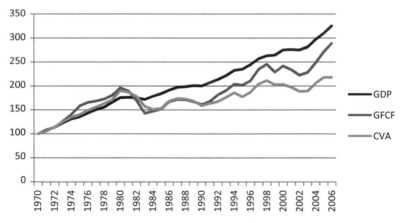

Figure 5.11 Volume indices of GDP, GFCF, CVA at constant 1990 US$ (1970 = 100) (South America and Caribbean).

Concluding remarks

The picture that emerges from the analysis suggests that the share of construction in gross output tends to increase with the level of per capita income in the first stages of economic development. When countries reach a certain level of economic development, the construction output will grow slower than the GDP in the latest stages of their recovery. That is, it decreases relatively but

Table 5.8 Value of GDP, GFCF and CVA and share of CVA in GDP in the world (constant 1990 US$ billions)

Year	1970	1971	1972	1973	1974	1975	1976	1977	1978	1979
GDP	11,502	11,970	12,642	13,432	13,699	13,837	14,538	15,119	15,770	16,388
GFCF	2,713.0	2,858.8	3,040.4	3,289.5	3,237.7	3,219.33	3,379.3	3,521.3	3,717.0	3,884.3
CVA	961.7	987.4	1,037.9	1,075.8	1,054.4	1,066.0	1,101.2	1,113.3	1,161.0	1,182.2
CVA/GDP (%)	8.37	8.25	8.22	8.00	7.70	7.70	7.58	7.37	7.37	7.21

Year	1980	1981	1982	1983	1984	1985	1986	1987	1988	1989
GDP	16,708	17,029	17,169	17,650	18,429	19,074	19,679	20,356	21,300	22,122
GFCF	3,881.7	3,887.8	3,815.6	3,856.0	4,023.3	4,182.9	4,381.9	4,504.2	4,810.0	5,035.0
CVA	1,146.5	1,130.5	1,112.1	1,097.0	1,102.7	1,136.3	1,175.0	1,221.7	1,310.0	1,385.1
CVA/GDP (%)	6.86	6.63	6.48	6.21	5.98	5.96	5.97	6.00	6.15	6.26

Year	1990	1991	1992	1993	1994	1995	1996	1997	1998	1999
GDP	22,130	22,440	22,872	23,217	23,963	24,663	25,464	26,385	27,021	27,905
GFCF	5,038.2	5,011.1	5,046.5	5087.1	5,299.4	5,535.8	5,842.5	6,116.9	6,302.3	6,562.9
CVA	1,340.9	1,311.2	1,301.1	1,299.6	1,320.6	1,310.6	1,334.9	1,350.1	1,367.1	1,377.4
CVA/GDP (%)	6.06	5.84	5.69	5.60	5.51	5.31	5.24	5.11	5.06	4.94

Year	2000	2001	2002	2003	2004	2005	2006
GDP	29,066	29,564	30,133	30,972	32,209	33,336	34,694
GFCF	6,882.0	6,888.4	6,861.5	7,130.8	7,539.1	8,084.0	8,597.8
CVA	1,389.9	1,403.8	1,410.6	1,433.8	1,475.6	1,517.7	1,578.0
CVA/GDP (%)	4.78	4.75	4.69	4.63	4.58	4.55	4.55

Table 5.9 Construction value added as a share of GDP in Group I (mean)

	Year									
	1970	1971	1972	1973	1974	1975	1976	1977	1978	1979
CVA/GDP (%)	3.89	3.91	3.93	3.94	3.94	3.99	4.12	4.14	4.11	4.31
	Year									
	1980	1981	1982	1983	1984	1985	1986	1987	1988	1989
CVA/GDP (%)	4.41	4.38	4.38	4.34	4.29	4.11	4.13	4.28	4.27	4.26
	Year									
	1990	1991	1992	1993	1994	1995	1996	1997	1998	1999
CVA/GDP (%)	4.29	4.36	4.61	4.52	4.47	4.51	4.64	4.57	4.47	4.53
	Year									
	2000	2001	2002	2003	2004	2005	2006			
CVA/GDP (%)	4.69	4.73	4.90	5.01	5.07	5.17	5.19			

Table 5.10 Construction value added as a share of GDP in Group II (mean)

	Year									
	1970	1971	1972	1973	1974	1975	1976	1977	1978	1979
CVA/GDP (%)	6.56	6.60	6.57	6.58	6.62	6.81	7.13	6.89	6.90	6.93
	Year									
	1980	1981	1982	1983	1984	1985	1986	1987	1988	1989
CVA/GDP (%)	6.94	7.06	6.99	6.78	6.44	6.10	5.94	5.77	5.66	5.67
	Year									
	1990	1991	1992	1993	1994	1995	1996	1997	1998	1999
CVA/GDP (%)	5.58	5.55	5.68	5.75	5.77	5.78	5.70	5.56	5.55	5.51
	Year									
	2000	2001	2002	2003	2004	2005	2006			
CVA/GDP (%)	5.52	5.60	5.58	5.44	5.48	5.62	5.53			

Table 5.11 Construction value added as a share of GDP in Group III (mean)

Year									
1970	1971	1972	1973	1974	1975	1976	1977	1978	1979
CVA/GDP (%) 8.25	8.23	8.28	8.04	7.81	7.83	7.58	7.50	7.37	7.18

Year									
1980	1981	1982	1983	1984	1985	1986	1987	1988	1989
CVA/GDP (%) 6.86	6.70	6.65	6.58	6.41	6.15	6.13	6.06	6.11	6.12

Year									
1990	1991	1992	1993	1994	1995	1996	1997	1998	1999
CVA/GDP (%) 6.06	6.02	5.94	5.67	5.71	5.55	5.65	5.77	5.70	5.58

Year							
2000	2001	2002	2003	2004	2005	2006	
CVA/GDP (%) 5.44	5.38	5,31	5.43	5.32	5.41	5.23	

not absolutely. Thus, it is reasonable to assume that when a certain level is achieved (say a share of CVA in GDP of around 5–6 per cent of GDP – it depends upon the year taken as base and the currency used) and countries enter into a period of sustained economic growth and development, the construction output tends to grow, in general, with the same rate of growth of that of the GDP.

The construction sector plays an important role in the development strategy of any country that goes beyond its share in national output. Many writers have referred to its effect on employment creation, others to its multiplier effects in the national economy. Another relevant feature should be added: it is the great flexibility of the construction industry activity in adjusting to different framework conditions that makes, particularly, this sector of the economy a major contributor to the process of economic growth and development.

References

Barro, R.J. (1991) Economic growth in a cross section of countries. *The Quarterly Journal of Economics*, **106**, 407–443.

Bon, R. (1990) The world building market, 1970–1985. *Proceedings of the CIB International Symposium on Building Economics and Construction Management*, **1**, Sydney, March 14–21, pp. 16–47

Bon, R. (1992) The future of international construction: secular patterns of growth and decline. *Habitat International*, **16**(3), 119–128.

Chenery, H.B., Robinson, S. and Syrquin, M. (1986) *Industrialisation and Growth: A Comparative Study*. Published for the World Bank, Oxford University Press: Oxford.

De Long, J.B. and Summers L.H. (1991) Equipment investment and economic growth. *The Quarterly Journal of Economics*, **106**, 445–502.

Denison, E. (1985) *Trends in American Economic Growth, 1929–1982*. The Brookings Institution: Washington, D.C.

Hickman, B.G. and Choen, R.M. (1987) *An Annual Growth Model of The U.S. Economy*. Contributions to Economic Analysis, No. 100. North-Holland: Amsterdam.

Kuznets, S. (1968) *Towards a Theory of Economic Growth*. W.W. Norton: New York.

Lewis, W.A. (1955) *The Theory of Economic Growth*. Allen & Unwin: London.

Lopes, J. (1997) *Interdependence between the Construction Sector and the National Economy in Developing Countries: A Special Focus on Angola and Mozambique*. Unpublished Ph.D. Thesis, University of Salford, UK.

Lucas, R. (1988) On the Mechanics of Economic Development. *Journal of Monetary Economics*, **22**, 3–42.

Maddison, A. (1987) Growth and slowdown in advanced capitalist economies. *Journal of Economic Literature*, **25**(2), 649–698.

Maddison, A. (2006) *The World Economy: A Millennial Perspective*. OECD: Paris.

Olson, M. (1996) Distinguished Lecture on Economics in Government: Big bills left on the sidewalk; why some nations are rich, and others poor. *Journal of Economic Perspectives*, **10**(2), 3–24.

Reynolds, L.G. (1985) *Economic Growth in the Third World: 1850–1980*. Yale University Press: New Haven.

Romer, P.M. (1990) Endogenous technological change. *Journal of Political Economy*, **98**(5), S71–S102.

Ruddock, L. (2000) *An International Survey of Macroeconomic and Market Information on the Construction Sector: Issues of Availability and Reliability, RICS Research Papers*, Vol. 3, No. 11. RICS Research Foundation: London.

Ruddock, L. and Lopes, J. (2006) construction and economic development: the Bon Curve. *Construction Management and Economics*, **24**, 717–723.

Solow, R. (1956) A contribution to the theory of economic growth. *Quarterly Journal of Economics*, **70**, 65–94.

Todaro, M.P. (1992) *Economic Development in the Third World*, 3rd edition. Longman Group Ltd.: New York.

Turin, D.A. (1973) *The Construction Industry: Its Economic Significance and its Role in Development*. UCERG: London.

United Nations (2008 Internet edition) *Yearbook of National Accounts Statistics*. Statistical Office, DESA: New York.

Wells, J. (1986) *The Construction Industry in Developing Countries: Alternative Strategies for Development*. Croom Helm Ltd: London.

World Bank (1984) *The Construction Industry: Issues and Strategies in Developing Countries*. IBRD, The World Bank: Washington, D.C.

World Bank (1986, 1992, 1994) *World Development Report*. IBRD, Oxford University Press.

Chapter 6

The impact of fiscal, monetary and regulatory policy on the construction industry

Geoff Briscoe

Introduction

This chapter examines the most significant effects that changes in government fiscal, monetary and regulatory policies exert on the construction industry. In order to understand why governments make on-going adjustments to these different policies, it is necessary to explore the key macroeconomic objectives that most present-day governments seek to achieve. In the next section, these core objectives are examined: economic growth, full employment, control of inflation, stability in the balance of payments and the external exchange rate and the protection of the environment. When many aspects of these policies are changed, there will often be a significant impact on the construction sector. The areas of fiscal, monetary and regulatory policy are separately looked at in detail.

- The aims and scope of fiscal policy are described and the main sources of revenue and the applications of government expenditures are identified. Increased levels of taxation in an economy constrain the ability of both households and companies to spend on construction products and services, but increased levels of public sector spending will benefit the construction sector, particularly when this expenditure is directed towards capital projects.
- The objectives of monetary policy and the tools that it employs are explained. Whilst adjustments to interest rates are the most conspicuous monetary control technique, other measures that have a less direct, but significant, impact on construction are identified. When monetary policy is tightened, so that the cost of credit is raised and bank advances become more restrictive, both the construction companies and many of their clients are adversely affected.
- The main types of regulatory policy that governments use to protect the environment are summarized. Building regulations, environmental directives and general planning legislation all constrain companies in the design and implementation of construction projects. Whilst this chapter makes

reference to the execution of fiscal, monetary and regulatory policies in a number of countries and especially in the member states of the European Union, it predominantly describes how these polices are enacted in the United Kingdom.

Government macroeconomic policy objectives

There is a wide consensus amongst modern governments as to their common macroeconomic objectives: a sustained rate of long-term economic growth, acceptably low levels of unemployment, price and wage stability and a trade balance with other countries that leads to exchange rate stability. These are the enduring primary objectives and, increasingly, governments have added to this list the objective of protection of the environment. There exist many other secondary objectives, but these tend not to be the prime concern of fiscal and monetary policy changes.

Economic growth is measured by the annual rate of change of real gross national product (GNP) per capita. Sometimes, gross domestic product (GDP) is used instead of GNP and often, for purposes of international comparison, growth is revalued to take account of purchasing power parities. Successful economies in the EU manage to achieve average growth rates of between 2 and 3 per cent p.a. Some developing economies achieve much higher rates than these; China has in recent years (2006–7) been experiencing growth rates in excess of 10 per cent, whilst India has been growing at some 7 per cent p.a. It is economic growth that produces higher living standards and more wealth for the nation. Higher levels of growth are usually accompanied by increased levels of fixed investment in dwellings, buildings and infrastructure. Construction activity benefits significantly from strong economic growth in a country, whereas weak growth or even annual declines in real GNP can lead to recession in the construction sector, such as occurred in the early 1990s in the UK.

Pursuit of low levels of unemployment across the national workforce is another core macroeconomic objective. It is extremely wasteful of scarce labour resources to have significant numbers of the nation's workforce unemployed. Most EU governments become concerned when registered unemployment begins to rise significantly above about 6 per cent of those actively seeking work. It is widely acknowledged that even in a situation of relatively full employment, there will always be some workers who are between jobs and others who are retraining to new jobs. The accurate measurement of unemployment is rendered difficult by issues of part-time and temporary jobs and also by those people who are nominally part of the working population but are not actively seeking work because they are registered disabled, in full-time education or training, or they have opted for early retirement. When governments initiate large-scale capital expenditure programmes with an intention to create employment opportunities they will, typically, have high construction content.

It is widely accepted that control of price inflation is a primary government objective, as high rates of annual price increases are considered damaging to most sectors of the economy. In particular, relatively high price inflation in a nation will exert an adverse impact on its balance of trade, as the relative price of its exports rise and its import prices fall. Elsewhere, inflation is found to have a distorting impact on the distribution of income across the population; those in employment can expect wage levels to keep pace with, or even outstrip, rising prices, whereas others, dependent on state benefits and fixed pensions, may experience reductions in real incomes. Inflation also often creates marked price discrepancies in different goods and service markets; in the UK over recent decades the price of housing has increased much more rapidly than average retail prices. It was once thought that high inflation was acceptable if it led to higher levels of employment, but today the argument is for stable prices, required for business confidence leading to the higher fixed investment that is necessary to create more jobs. Governments throughout the EU seek to restrict annual inflation rates, as measured by the consumer price index, to below 3 per cent. Where alternative measures of price inflation, such as the retail price index, are used, different target rates may apply. More stable prices are beneficial for the construction sector as they serve to contain tender prices and they produce an increased likelihood of client's cost budgets being met.

Over a number of years, a nation aims to ensure that the amount that it spends and invests abroad more or less balances with the amounts other countries spend and invest in it. Overall, these external transactions will balance out, as any deficit on the current account (trade in visible goods and invisible services) is accommodated by positive inflows on the capital account (net investment and official financing flows). Where a nation continues to run a large current account deficit, it will need to generate increasingly large capital inflows from foreign investors or accept progressive reductions in its official national reserves. The US in recent years has incurred relatively high current account deficits that are mainly funded by foreign nationals (e.g. China) investing their current account surpluses back into US companies and US bank deposits. Persistent and significant deficit balances on the current account are an important factor influencing the strength of that country's foreign exchange rate. Usually, unless steps are taken to correct the deficit, it will engender a downward pressure on the exchange rate, where that currency is free floating. For those countries that are joined in the European Monetary Union (EMU), the strength of the euro is determined by their collective trade balances. Construction organizations who carry out work overseas and import significant volumes of materials and components benefit from stable exchange rates and they may be adversely effected by short-term fluctuations in these rates.

More recently, many governments, especially those in the EU, have added protection of the environment to their set of key macroeconomic objectives. The main thrust of the objective is the reduction of carbon (CO_2) output that is the prime cause of global warming. Other aspects are concerned with the reduction of pollution in its different forms, the minimization of waste, the preservation of biodiversity and the establishment of sustainable development practices. Economic decision-making in private-sector markets typically fails to take into account the external social costs arising from the collective actions of both firms and individuals, so that government must rectify this market failure if environmental targets are to be met. Policies to deal with environmental issues are likely to have a strong impact on construction, as buildings and dwellings are significant generators of carbon and firms in the industry are often implicated in poor waste management and pollution practices.

Fiscal policy

Fiscal policy is concerned with the general government sector budget that relates to taxation and other sources of government revenue, expenditure on current, capital and transfer items and any borrowing requirement that might arise. In the UK, fiscal policy is controlled by HM Treasury and the budget is managed according to two key principles or rules: over the economic cycle public sector debt, expressed as a proportion of GNP, is required to be held stable and at a prudent level (usually no more than 3 per cent of GNP), and government can only borrow to carry out investment and not for the purpose of current spending. Across the EU national governments are free to set their own fiscal policies, although for those countries that are members of the EMU central guidance on setting fiscal budgets is provided and rules similar to those in operation in the UK are adopted. The construction sector is very significantly affected by fiscal changes, especially in regard to a range of tax adjustments and to government spending allocations on capital projects.

Aims of fiscal policy

The prime aim of fiscal policy is to stabilize the economy to enable the core macroeconomic objectives to be realized. In particular, fiscal policy is expected to have its main impact on economic growth and reducing unemployment. Annual government budgetary adjustments are made to fine-tune the economy and to try to ensure that a balance is maintained between consumer and investment expenditures. Governments often make on-going changes to their fiscal plans between budgets.

Fiscal policy is also used to ensure that the national GNP is distributed fairly across the individuals who make up the population. Governments do not necessarily aim to achieve more equitable distributions of wealth or income, but they are concerned to financially protect the poorest groups in society. In this context, tax adjustments are made to collect revenue from those on higher incomes and make transfer payments to those on lower incomes. In this way, the poorer groups are enabled to spend on consumer goods and services, including housing services.

Fiscal policy is applied to provide the public goods and services in an economy that private markets normally will not provide. Usually, this means the government sector paying for education, health, defence, transport, law and order and similar public services. The fiscal budget must accommodate both the capital projects (incorporating buildings and infrastructure) and the current spending to provide the wages and salaries of those who work in the public sector. These public goods and services will need to be paid for out of taxes and government revenues. Each national government needs to determine where to draw the line on what services the public sector provides and which services can be accommodated by the private market. In the UK over the period since 1980 a wide range of transport, utility and housing services, previously located in the government sector, have been privatized and transferred into the control of commercial organizations. More recently, the UK government has established the Private Finance Initiative (PFI) method of funding many of its construction projects. This change eases the short-term burden on fiscal capital expenditures but transfers risks and management responsibilities to construction companies and their financial partners.

Increasingly, governments are using fiscal policy to assist in realizing their environmental objectives. A range of taxes, such as the UK's carbon and landfill taxes, have been introduced to raise the market prices of goods and services that are deemed to exert a detrimental impact on the environment. In this way, both businesses and households are deterred, through the price effect, from continuing to engage in consumption patterns and practices that either add to carbon output or generate significant amounts of waste. Revenue raised from such taxation can be used to subsidize goods and services that are beneficial to the environment; an example is the payment of grants to house owners to enable them to insulate their properties to a higher standard.

Sources of revenue

Fiscal revenues are collated from a wide variety of sources, predominantly by central government, but also by local governments and other agencies that are part of the public sector. Table 6.1 details the various revenue sources for the UK fiscal budget for the accounting year 2007/8. These are planned estimates and the actual outturns may well show significant variance. The budget assumes certain levels of activity in the economy and should this level not be realized, actual tax revenues will fall short of the expected total.

The prime source of fiscal revenue in the UK (and in most other countries) is income tax and the associated national insurance contributions. Such tax is levied on gross incomes of all UK residents. Personal allowances mean that those on low incomes are exempt from tax, whilst there are a limited number of other allowances and reliefs that are available to reduce the individual's tax liability. There are two main tax bands in the UK, the basic rate (applicable to the majority of the working population) and the higher or surtax rate.

Table 6.1 UK fiscal revenues and expenditure 2007/8 estimates. (Adapted from HM Treasury 2007 Pre-Budget Report and Comprehensive Spending Review: Annex B: Public Finances.)

Revenues	£Bn	% total	Expenditures	£Bn	% total
Income tax	150	27.2	Social Security benefits	158	26.8
National insurance	97	17.6	Health (incl. NHS)	90	15.3
Corporation tax	47	8.5	Education & skills	78	13.2
Business rates	22	4.0	Local government & communities	55	9.3
Council tax	24	4.4	Regional government	51	8.7
VAT	81	14.7	Defence	40	6.8
Excise duty	41	7.4	Transport	21	3.6
Capital taxes (incl. stamp duties)	24	4.4	Home Office (incld. law & justice)	20	3.4
Other revenue (incl. LA rents)	65	11.8	Foreign & international development (incl EC transfers)	15	2.5
			Other spending	31	5.3
Govt borrowing	38	6.9	Debt interest	30	5.1
Total receipt	**551**	**100.0**	**Total expenditure**	**589**	**100.0**

HM Treasury annually makes small adjustments in both the personal allowances and the tax bands, so that the revenue collated from income tax can be changed in line with the fiscal requirements. Differential tax rates apply to earned income and income from savings and dividends. Theoretically, national insurance payments are separate from income tax (although both are collected by the same agency, HM Revenue and Customs), and these deductions from income fund state pensions, unemployment and housing benefits and various other transfer benefits payable to those on low incomes. Significantly, it is not only employees and, to a lesser extent, the self-employed who pay national insurance, but also employers pay a percentage of the gross wage into the national insurance fund when they directly employ a worker. Again, fiscal adjustments are made from time to time in the national insurance contribution rates.

Corporation tax is a tax on the profits of companies and similar organizations. Companies normally use a range of different allowances to offset their final tax liabilities and, where companies make a trading loss in any year, this loss can usually be carried forward to set against future tax liabilities. Many large organizations in the UK often pay relatively small amounts of corporation tax once they have claimed all the possible allowances. Rates of corporation tax vary within a country dependent on the level of earned profit, after allowances, and they also vary between countries. Table 6.2 shows the marked differences

Table 6.2 Corporation tax and VAT rates in the EU (standard rates applicable to larger companies in 2006/2007). (Source: KPMG International (www.kpmg.com).)

Country	Corporation tax rate (%)	VAT rate (%)
Austria	25	20
Belgium	34	21
Bulgaria	15	20
Cyprus	10	15
Czech Republic	24	19
Denmark	28	25
Finland	26	22
France	33.3	19.6
Finland	28	22
Germany	38.3	19
Greece	29	19
Hungary	16	20
Ireland	12.5	21
Italy	37.3	20
Latvia	15	18
Lithuania	15	18
Luxembourg	30	15
Malta	35	18
Netherlands	29.6	19
Poland	19	22
Portugal	27.5	21
Romania	16	19
Slovakia	19	19
Slovenia	25	20
Spain	35	16
Sweden	28	25
United Kingdom	29	17.5
EU average (27 countries)	**25.2**	**19.6**

between the EU nations. Most of the newer countries that joined the EU in 2004 and subsequently have comparatively low rates of corporation tax used to attract foreign companies to become inward investors. Several longer-standing members of the EU have recently cut their corporation tax rates to compete for this inward investment. Ireland is the outstanding example of how a lowering of corporation tax has induced foreign firms to relocate there to reduce their tax bills.

Value added tax (VAT) is an important revenue raising tax that is levied on consumer expenditures. In recent years, many governments and especially the UK have increased the amount of revenue collected from this tax by extending the range of goods and services to which the tax is applied and, in some cases, by increasing the rate of the tax. Table 6.2 shows how there is no current uniformity in the application of VAT across the EU nations, although many EU policy makers have argued the case for greater harmonization. VAT is added to the final price of goods and services that consumers pay and differential rates of tax can lead to price distortions between countries. In the UK, no VAT is levied on new housing but in Sweden a standard rate of 25% is applied. Expenditure taxes, unlike income taxes, impact equally on both rich and poor members of society at the point of purchase. Companies are able to reclaim any VAT that they pay on purchased materials and services, but they are the collating agents for the government and there is a significant expense in the administration of the VAT revenues.

There remains a number of other sources of government revenue, some of which exercise an impact on the construction sector. Companies pay business rates and householders pay local council taxes on the properties that they occupy. Usually the larger the property, the more tax is payable, although council taxes are locally determined and geographic variations arise. In the UK, house and property transactions are subject to stamp duties, with rates increasing progressively as property values increase. The UK government applies an inheritance tax to the higher valued estates of deceased individuals and usually it is the high market value of the house that increasingly draws the estate into the band where taxes become payable.

Patterns of expenditure

Governments need to plan their expenditures so that the aggregate spending matches as closely as possible the expected aggregate revenues. Where planned expenditure exceeds revenue, the government will need to borrow to fund the deficit budget. Usually, the total expenditure needs far exceed the revenues that are available to fund the various programmes and the Treasury department has to make some budgetary choices. The majority of government spending is applied to meet current expenditures in respect of transfer benefits and wages and salaries for those who work in the public sector. The smaller residual component of the expenditure budget is allocated to capital programmes, and the construction industry is a particular beneficiary of such spending. Where decisions are made to increase spending in one policy area (e.g. defence), then in the absence of any marked increase in revenues, or borrowing for capital projects, reduced spending may well result in other areas (e.g. social housing).

Table 6.1 shows the distribution of government spending for the UK. The largest expenditure component is the payments made for a wide range of social security benefits, including state pensions, job-seekers allowances and, particularly important for construction, housing benefit. Without this housing subsidy, many of the tenants in the rented dwellings sector would not be able to meet their rent commitments to local authorities, housing associations and private landlords. This fiscal benefit acts as a current account transfer payment that enables the providers of rented housing to continue with their investment programmes.

Other major areas of government expenditure are the funding of the National Health Service (NHS) and a wide range of public educational services. These two sectors account for more than a third of UK government expenditure and this spending has been growing significantly in recent years. Whilst the majority of this expenditure is on wages and salaries, increasing amounts are used to pay for annual rental charges on new buildings provided under the PFI. The expenditure budgets of other departments are relatively small compared to social security benefits, NHS and education, but some of these others are critically important for the construction industry. The public sector expenditure on housing, the environment and transport creates a lot of work for contractors, consultants and public authority construction staff.

Main impacts on construction

Table 6.1 shows how some 45 per cent of UK fiscal revenue arises from combined income tax and national insurance. These taxes determine how much disposable income individuals will have to spend on consumer goods and services, including housing and construction-related services. When real disposable incomes are increasing, demand for housing will also increase. Conversely, income tax increases (in whatever form) will exert a negative impact on this demand and this is especially the case for first-time buyers who are already likely to have below-average incomes. Corporation taxes reduce the amount of profit private sector companies are able to retain and apply towards new investment in buildings and related capital investments; companies increasingly favour location in countries with lower regimes of corporation tax. VAT serves to increase the final price of a product or service to the end-consumer and may well dampen demand. In the UK, for many years now, VAT has been applied across all materials and services used in construction repair and maintenance and it is also charged on the construction of private sector industrial and commercial new work. Some small companies who are not registered for VAT are able to carry out construction contracts at a lower price than their larger counterparts who must make VAT returns.

Capital expenditure by the public sector client is very important to many parts of the construction industry. In the UK, the distinction between current and capital expenditures has become blurred by the government preference of funding most new capital programmes and large refurbishment projects through the PFI and paying for the resulting buildings and structures by annual rentals over a number of future years. Smaller-scale repair and maintenance work on the public sector building stock is usually funded by traditional direct payments. In total, the various government departments annually put forward construction programmes that cannot be simultaneously funded without violating the fiscal budget. Rolling programmes of construction work are drawn up in the expectation of future funding when the expenditure budget will allow.

Monetary policy

Monetary policy is usually controlled by banking authorities on behalf of
national governments, the Bank of England (BoE) for the UK and the European
Central Bank (ECB) for those countries who are members of the EMU.
Monetary policy is primarily concerned with determining the availability and
price of credit in the economy. Much consumer and investment spending is
carried out using borrowed money and, by setting the baseline terms and
conditions on which money can be borrowed, the BoE, ECB and similar
authorities elsewhere aim to strongly influence planned demand expenditures.
Construction is very dependent on loans from the finance sector to fund
client demand for buildings and dwellings, as well as to provide the credit
needed by companies and consultancies to meet cashflow requirements in
fulfilling their contracts. Elsewhere, monetary policy is concerned with the
regulation of external exchange rates. Construction is an international industry that uses extensive imports of materials and it also often exports its
services in carrying out overseas construction projects. The exchange rate and
its stability is an important determinant of these activities. Table 6.3 summarizes the main groups of control measures that collectively constitute monetary policy.

Aims of monetary policy

Monetary policy is used primarily to control inflation in the economy. Interest
rates are raised and money supply is tightened when inflationary pressures start
to increase and these measures are reversed when it is deemed prudent to boost
spending, without creating significant price and wage acceleration. Unlike fiscal

Table 6.3 Key components of UK monetary policy

1	**Changes in bank rate by the Bank of England (BoE)** This leads other banks and building societies to alter the interest rates that they charge on loans
2	**Regulation of the financial markets by the BoE** Licensing of banks and similar financial institutions, and setting the legal framework for their operating conditions. The BoE acts as lender of the last resort, providing support for failing banks
3	**Monitoring of money supply to the economy by BoE** Loose control of the total amount of deposits held by the financial sector. Use of open market bond sales and purchases to change the amount of bank deposits in the economic system and so regulate the money supply
4	**BoE influence over the level of the sterling exchange rate** BoE is able to intervene in foreign exchange markets, either buying sterling to support its price or selling sterling to prevent its price being pushed too high. In this way, short-term fluctuations can be stabilized

change that can only be introduced with significant time lags, interest rates can be adjusted instantaneously, although the derived impact of such change may be slow to feed through into spending decisions. During the first half of 2007, the BoE raised its base rate upwards (in quarter percentage steps) in an attempt to control growing inflation in the UK economy; the resulting reduction in consumer spending only took effect several months later.

Another aspect of monetary policy is the monitoring of the currency exchange rate, such as the value of the £ sterling against other world currencies. The central bank in the economic system is able to buy and sell large amounts of the domestic currency in an attempt to smooth out short-term fluctuations in the exchange rate. In the long run, the bank's ability to regulate the rate is likely to prove more limited as trade and capital flows become the key determinants.

Interest rate changes

The prime instrument in effecting monetary policy is changes in interest rates. When central banks make adjustments to the base rate at which they are prepared to lend to other banks and financial intermediaries in the economy, this has a related impact on the interest rates that the commercial banks charge their customers. If the BoE raises its base rate this signals to other banks in the economic system that, should they need to approach the BoE for funds to maintain their liquidity position, they will now be obliged to pay at the higher rate. Usually, this results in the commercial banks adjusting their own loan rates upwards by at least the same amount as the increase in bank rate. Conversely, if the central bank is seeking to lower the cost of borrowing across the economy and so facilitate higher levels of consuming spending out of credit, it will lower its bank rate and signal to the other financial institutions that they should lower their interest charges. There is no obligation on the part of banks to immediately follow these signals from the central bank, although they usually will adjust their rates (sometimes with a significant delay) in order to maintain their competitive position.

The interest rates that commercial banks charge their borrowers are normally at a significantly higher level than bank rate. Banks take account of the credit ratings of potential applicants for loans and the interest rates that they charge reflect the perceived degree of risk in any default on repaying the loan. The lowest-risk customers are able to borrow at rates close to or only a percentage point or so above bank rate; many larger construction organizations, with sound order books, fall into this category. Higher-risk customers are charged much higher rates of interest well above bank rate, especially on short-term overdraft loans; most small construction firms and sole traders fall into this group. In setting their loan rates banks take account of the need to attract an adequate inflow of deposits and savings, so that savings rates will also be adjusted when bank rate is changed. Increased savings rates will lead to higher loan rates where the banks seek to maintain or improve profit margins.

Lending by banks

The central bank has other instruments, apart from bank rate, that it can implement to effect monetary policy. The central bank is generally responsible for the regulation of the wider banking and financial system in the economy. In the UK this responsibility is shared between the BoE and the Financial Services Authority (FSA). Regulation involves licensing of approved financial institutions and monitoring to ensure that such banks operate prudently to maintain safe liquidity ratios. Occasionally, the central bank is obliged to intervene to rescue a commercial bank which encounters cash flow problems. A recent example of this took place in September 2007 when the Northern Rock bank in the UK got into short-term difficulties, as it became unable to raise the necessary funds from the inter-bank market to meet routine customer withdrawals. The BoE stepped in to exercise its role as 'Lender of the Last Resort' and so prevent a major bank failure that could have led to a general loss of confidence and possibly a wider financial panic. Whilst the central bank has a responsibility to maintain stability and overall confidence in the banking system, it is not obliged to rescue all banks and financial institutions that fail through poor management; in 1991 the BoE failed to provide a lifeboat operation to save the large Bank of Commercial Credit International (BCCI), where fraud was a prominent factor in its failure.

Banking is an international activity and monetary difficulties in one country can exercise a significant impact on bank lending policy in other countries around the world. The collapse in the US sub-prime loans market in 2007/8 led to monetary changes in many other countries, including the UK. Banks became more cautious in their lending policies as the risk of loan defaults was made more apparent. This resulted in higher-risk customers (including most first-time housing buyers) facing greater obstacles in obtaining loans and higher rates of interest being charged on the loans. The construction industry is adversely affected by such changes.

Other monetary measures

When the BoE seeks to increase the money supply to the economy, it is able to print large quantities of new banknotes and use them to purchase government securities (bonds of different kinds) currently held by either commercial banks or the private sector. The result of this so-called 'Open Market Operation' is to increase the monetary base of the UK banking system, so that when this money finds its way into bank deposits (as most of it is guaranteed to do so), the banks are now able to increase their loans and advances, on the basis of their improved cash reserves. The converse operation is where the BoE chooses to sell large amounts of government securities from its existing stock and so draw cash from either private circulation or directly from the cash reserves of the banks. Now the monetary base is reduced, the banks lose cash reserves and they are obliged to restrict deposit lending and so cause the money supply to fall. Where the fiscal budget produces a large deficit, the central bank will need to sell securities to effect the borrowings on behalf of the Treasury. The commercial banks can be expected to seek to avoid a significant reduction in their cash reserves and, by implication, a decrease in their profit-generating deposits. This can be achieved by increasing some of their interest rates to attract new saving inflows and higher loan interest rates will result.

Another instrument at the disposal of the central bank is the ability to carry out short-term interventions in the international money market where the exchange rate is determined. It has already been explained how, over the long run, changes in the balance of payments are the prime determinant of a nation's currency standing relative to other world currencies. However, exchange rates can be subject to short-term volatility as speculators buy and sell large volumes of particular currencies in anticipation of changes. Central banks can use their extensive financial resources to buy in a domestic currency that is increasingly being sold short on the international monetary exchanges in expectation of price deterioration. By purchasing all of the national currency that is sold at a fixed price the central bank can usually prevent large-scale falls in the exchange rate. The limitations of the BoE to withstand speculative selling pressure on the £ sterling were exposed in 1992 when sterling was forced to exit from the Exchange Rate Mechanism (a predecessor of the EMU), with a resulting large downward revision against other key European currencies. The central bank may also want to intervene to prevent its currency rising too rapidly against other world currencies. Here the central bank would sell large volumes of its currency at the chosen exchange rate to satisfy the speculative buying pressures. These powers of intervention are a short-term instrument to try to ensure greater stability in exchange rates.

Main impacts on construction

Monetary policy changes have a strong impact on a construction industry that has a high dependency on finance and credit. Most construction companies operate bank overdrafts which are essential for ensuring cash flow on projects where payments are often made for materials, plant, direct labour, design and subcontract services before revenue is received from clients for work done. Housebuilding firms commonly build speculatively and they only obtain revenue upon completion and sale. Larger contractors and consultancies enter into PFI ventures where all the construction work must be fully funded before either a stream of rental revenues is received or the financial interest in the completed venture can be sold on. Property developers also operate in a similar way, using significant amounts of bank finance to pay for the project before it can be rented or sold to an institutional investor. The construction companies need to secure bank finance for both short- and long-term purposes and the interest rates they have to pay on the loans is an important factor in determining the profitability potential of the individual project and, often, the overall business.

It is not only the construction companies for whom finance can be critical but also the clients who initiate the projects and those who must eventually pay for the completed buildings and structures. Most new dwellings in the UK are purchased using mortgage advances from banks and building societies. Where the BoE carries out open market operations to reduce the money supply, this may cause the lending institutions to restrict the number of advances they can provide in a given period and some higher-risk borrowers especially first-time buyers, may be forced to delay their housing purchases. When bank rate is raised progressively over a short time period (as occurred in the UK in 2007) it will lead to significant increases in the mortgage rate and some potential buyers may no longer be able to afford to enter the owner occupation market. Elsewhere, the private commercial and industrial clients for buildings and works may be forced to reconsider the financial viability of proposed projects in the light of both higher borrowing costs and expected increased building costs, where the contractors seek to recover their higher finance charges.

Larger construction companies carry out projects overseas and many of these same organizations import large volumes of materials from around the world. The profitability of such activity will be sensitive to movements in the international exchange rates. Slow adjustments, such as those that typically take place between the major world currencies, pose only mild concern to those organizations involved in international contracting. Where the central banks fail to control sharp fluctuations in other exchange rates, much uncertainty can develop and international activity in construction markets may reduce as a result. In 2007, UK contractors and consultants working in the US experienced a gradual erosion in their profits as a result of the £ sterling slowly strengthening against the US $; in this situation contracts valued in the $ currency are worth less to UK companies. Equally, imported materials, such as timber which is usually priced in US $, will be cheaper in the UK as result of this currency trend. Monetary policy is important for ensuring the currency stability that is needed as the basis for international trade.

Regulatory policies for construction

The construction industry impacts on both the economy and the environment and, as such, it is subject to a potentially wide range of regulatory policies. This chapter does not cover wider economic regulation, such as competition policy or labour market legislation; rather the present concern is with those specific policies aimed to protect the environment from possible damage caused by both construction process and products. The most direct type of environmental control is that associated with the building regulations. Most countries seek to regulate the building process to ensure that minimal standards are implemented; the UK imposes some very strict standards. Another important group of regulatory policies arise from the various Economic Commission (EC) directives on environmental issues. Where these principles are carried through into national regulation policies, their impact on the construction industry is very significant. Construction in all countries is usually subject to some form of planning and development controls and these controls serve to further regulate building activity. Regulatory policies can differ significantly between countries, including across the EU nations; this section refers mainly to UK regulation. The UK has recently established a Construction Industry Policy and European Regulatory Group (CIPER), whose brief is to examine the impact of new regulations on the construction industry.

The building regulations

These regulations specify a set of minimum standards to which all new buildings are required to conform and compliance is checked by local building control inspectors. The standards are subject to periodic revision and, in recent years, UK energy regulations, in particular, have been revised very significantly to comply with higher environmental requirements. The building regulations cover almost all aspects of buildings and they constitute a very detailed set of standards.

Table 6.4 details the main subsections of the UK building regulations. It can be seen how the coverage is very far reaching, from structural stability, to internal services, to access standards. The regulations often need revisions to take account of new materials and components in use. These regulations will usually seek to incorporate any new EC directives, although often there is a significant time lag. Legislative changes, developed elsewhere in the economy but which contain building implications (e.g. disability discrimination), will be

Table 6.4 Summary of the UK building regulations (2007 approved UK building regulations)

Part A	Structure
Part B	Fire safety
Part C	Site preparation and resistance to moisture
Part D	Toxic substances
Part E	Sound insulation
Part F	Ventilation
Part G	Hygiene
Part H	Drainage and waste disposal
Part J	Combustion appliances and fuel storage
Part K	Protection from falling, collision and impact
Part L	Conservation of fuel and power
Part M	Disabled access to and use of buildings
Part N	Glazing
Part P	Electrical safety
Regulation 7	Materials and workmanship

accommodated in the building regulations. Recently, there have been significant changes made to Part L, as the regulations seek to impose greater efficiency in the use of energy and to reduce the carbon emissions from new dwellings and buildings. Further revisions are likely as the UK government strives to meet demanding EU targets for carbon reduction.

EC directives

The EC produces a large number of directives that provide policy guidance to national governments on a wide range of issues, many of which impact on construction. This is a form of indirect regulation that becomes direct when it is implemented into national law. The Construction Products Directive has been in existence for several years now and this determines specification standards for materials and components across the EU. The European Procurement Directive is concerned with ensuring competitive bidding procedures for larger construction projects.

Many of these directives are aimed at aspects of the environment. There are EC directives on water quality, air quality and noise control. There are directives that recommend the use of both environmental impact and strategic environmental assessments. Each of these is important when larger construction projects are seeking planning permissions.

EC directives on controlling construction waste and related pollution have resulted in some strong regulatory control measures in the UK. Specific taxes (part of the fiscal revenues) have been introduced for the disposal of active waste in landfill sites. Such sites are licensed and strictly controlled, and fly-tipping of waste is prohibited, with large fines applied to contractors who fail to comply. Larger construction projects are required to produce detailed site waste management plans (SWMP); these plans must specify how waste is to be separated out and removed, with an emphasis on recycling as much construction waste as possible. High-profile construction projects, such as the London 2012 Olympics, are expected to show the best standards of waste management with only very small proportions of waste going to landfill sites.

General planning regulation

The construction sector in all economies is subject to some measure of planning control, as various government departments and agencies seek to regulate the development and use of land in the wider public interest. Whilst planning is applied primarily to ensure the delivery of key public goods and services, it has increasingly become a prime mechanism in achieving environmental objectives. In the UK, most forms of construction development (widely defined to include all changes to buildings, structures and land) are subject to planning control, although some categories of development, mainly those associated with the interiors of buildings, are excluded. Generally, all developers are required to apply for planning permission for their proposed construction works. In the case of major infrastructure projects, such as the M6 Toll and Heathrow Terminal 5, the process of obtaining this permission proved very lengthy and led to significant delays in commencing the construction work. The work involved in meeting planning requirements can significantly raise the cost of a construction project and it may, in some instances, prevent the proposed development from taking place. Planning regulations continually evolve and in the UK future changes are likely to be made to speed up the decision-making process and to ensure that all new construction development follows sustainable practices.

Conclusions

Governments make adjustments to their fiscal and monetary policies in an attempt to realize the core macroeconomic objectives. These changes often carry significant consequences for the construction industry. Fiscal tuning that involves changes in income and expenditure taxes impacts on both private sector consumer and investment spending on construction goods and services. Government expenditure is an important component of construction demand and changes here carry implications for sectors such as social housebuilding and infrastructure. Monetary policy adjustments have a less direct impact, but the cost and availability of loans and credit are strong determinants of both construction client demand and firm supply. Construction, mainly because of its potential impact on the wider environment, is an industry that is subject to some very specific regulatory policies. Changes in such policies, especially measures to ensure greater sustainability, often exert a significant impact on construction costs and profitability.

Recommended further reading

Artis, M. and Nixson, F. (2007) *The Economics of the European Union: Policy and Analysis*, Chapters 2, 11 and 12. Oxford University Press: Oxford.

Begg, D., Fischer, S. and Dornbusch, R. (2005) *Economics*, Part 4. McGraw-Hill: Maidenhead.

Billington, M.J., Bright, K.T. and Waters, J.R. (2007) *The Building Regulations: Explained and Illustrated*. Blackwell: Oxford.

Cullingworth, J.B. and Nadin, V. (2006) *Town and Country Planning in the UK*. Routledge: Oxon.

Eijffinger, S.C.W. and De Haan, J. (2000) *European Monetary and Fiscal Policy*. Oxford University Press: Oxford.

Myers, D. (2004) *Construction Economics: A New Approach*, Part C. Spon Press: London.

Sawyer, M. (2005) *The UK Economy*. Oxford University Press: Oxford.

Warren, M. (2000) *Economic Analysis for Property and Business*, Part 3. Butterworth-Heinemann: Oxford.

Chapter 7

From the short to the long term

History and development of leading indicators and building cycles

Johan Snyman

Introduction

The relationship between demand and price is a recurring theme in the literature on the built environment. It is appropriate, therefore, to include a chapter on fluctuations in demand levels and how to forecast them. To gain insight into the cyclical development of business cycles, the researcher should follow Keynes' famous dictum: the economist '. . . must study the present in the light of the past for the purposes of the future'. Of course, there are many ways of studying the future. In ancient times, kings consulted the Oracle at Delphi. In modern times we make use of such methods as time series analysis, econometric modelling, extrapolation, the Delphi technique, scenario planning, trend- and cross-impact analysis, probabilistic system dynamics, technological forecasting and early warning systems, to name a few. Time series analysis and econometric modelling are commonly found in developed countries as an aid to macro-economic analysis and forecasting. In this chapter, the primary focus will be on time series analysis and leading economic indicators. Two streams have developed in the search for useful leading economic indicators. The quantitative approach dominates research in the US. The qualitative approach (business survey method) prevails in Europe. Yet many economic forecasting bodies integrate quantitative and qualitative data quite usefully in short-term forecasting models.

This chapter is divided into four sections. First, it is appropriate to examine the meaning of business cycles. This is done as an overview, because a detailed analysis of business cycles (their history, causes, chronology and prognoses) would require a volume in itself. Second, the focus shifts to quantitative economic indicators that lead the business cycle. Third, qualitative indicators, as derived from business and consumer surveys, are examined. Essentially, these indicators are meant as an aid to short-term economic forecasting. Finally, we move *from the short to the long term* with an analysis of building cycles.

Business cycles

Fluctuations in levels of economic activity are quite common in developed countries. These movements are known as business cycles. A business cycle usually has four distinct phases: the upswing or recovery phase, the peak, the downswing phase and the trough. In the US, reference is often made to expansions or contractions in economic activity. Business cycles are said to be recurring, but not periodic (i.e. not perfectly regular). A distinction is often made between classical business cycles and growth cycles (Mintz, 1974; Zarnowitz, 1985, 1992; Boehm, 1990; Cloete, 1990; Klein, 1990; Mankiw, 1990). When gross domestic product declines in *absolute terms* from peak to trough, it is known as a classical cycle. Growth cycles are identified by variations in the *rate of growth* of gross domestic product. As economies developed rather rapidly in the post-war period, growth cycles predominated. The duration of business cycles normally varies from 3 to 7 years, measured from peak to peak or from trough to trough. Sometimes, however, they can last for a decade, as was the case in the US during the 1990s (Zarnowitz, 1999).

Quantitative leading economic indicators

Origins of quantitative leading indicators

Quantitative leading economic indicators form part of a system of leading, coinciding and lagging indicators developed by the National Bureau of Economic Research (NBER) in the US. This system is often referred to as the 'NBER methodology'. It consists of a scoring system for the assessment and selection of economic indicators. Sometimes, it is referred to as the 'business cycle indicators approach'. Development of this methodology can be traced back to the early works of writers such as Arthur F Burns and Wesley C Mitchell (1946). In his recent book, Alan Greenspan (2007: pp. 184–185) mentions that Burns was one of his predecessors at the US Federal Reserve and described him as 'an economist who did groundbreaking work on business cycles'. The initial work was followed up by Geoffrey Moore, Julius Shiskin, Victor Zarnowitz, Charlotte Boschan, Philip Klein, Kajal Lahiri, Anirvan Banerji and others who continued the research, further developing and refining the selection, scoring and compilation of leading economic indicators. This method of business cycle research, or variations on the theme, is currently being used in most developed countries for the analysis of economic conditions and prospects.

Fiedler has stated that the central idea of the indicator approach is to find a batch of statistical time series that conform well to the business cycle and show a consistent timing pattern as leading or coincident or lagging indicators. However, he warned that 'Such a description of the indicator approach to business cycle analysis is grossly oversimplified, almost a caricature – the sort of thing that led to the accusation that the cyclical indicators are "measurement without theory". In fact, the indicator approach is quite complex both in theory and practice' (Fiedler, 1990: p. 137).

The debate surrounding 'measurement without theory' was initiated by Tjalling C. Koopmans (who received a Nobel Prize for Economics in 1975 for his earlier work in the field of econometrics). In 1947 Koopmans reviewed the book published by Burns and Mitchell a year earlier entitled *Measuring Business Cycles*. He argued that the indicator approach was not based on a firm theoretical footing. Second, he asserted that this approach had severe limitations for policy purposes. His third point of criticism dealt with the presence of random variability in economic data. The debate was taken up by Vining in 1949, who defended the indicator approach. Auerbach (1982) addressed this issue in his article 'The Index of Leading Indicators: "Measurement without Theory", Thirty-five Years Later' and evaluated the cyclical behaviour of twelve individual time series, as well as the composite index of leading indicators. Auerbach (1982: p. 595) concluded that '. . . the index serves a useful purpose'. A few years later, Philip A. Klein and Geoffrey H. Moore (1985) pointed out that Mitchell's first major work on business cycles had been published in 1913 and that Mitchell had foreseen the necessity for theoretical explanations for periods of prosperity and depression. Unfortunately, Mitchell died before the theoretical volume that was to follow *Measuring Business Cycles* could be written. Klein (1983), Moore and Cullity (1993) and Oppenländer (1993) defended the indicator approach on separate occasions and concluded that the evidence demonstrates that the tendency for leading indicators to lead at business cycle turns has survived for at least 100 years and for more than 20 business cycles, in the case of the USA. Moore and Cullity (1993: p. 17) asserted that the NBER methodology had '. . . withstood the test of time and space, and had become one of the tools most widely used in short-term forecasting'.

NBER scoring methodology

As mentioned earlier, the seminal work on business cycles in the USA was undertaken by Burns and Mitchell in 1946. In the mid-1960s Moore and Shiskin reviewed the cyclical indicators. By the early 1970s an appraisal was deemed necessary because of rising inflation, the existence of growth cycles and economic stabilisation policies. In their study, Zarnowitz and Boschan (1975) applied six criteria, below, when assessing and selecting the cyclical indicators (with the contribution of each criterion in brackets):

• Economic significance: how well understood and how important is the role in business cycles of the variable represented by the data? (Score 16.7 per cent.)
• Statistical adequacy: how well does the given series measure the economic variable or process in question? (Score 16.7 per cent.)

- Timing at revivals and recessions: how consistently has the series led at the successive business cycle turns? (Score 26.7 per cent.)
- Conformity to historical business cycles: how regularly have the movements in the specific indicator reflected the expansions and contractions in the economy at large? (Score 16.7 per cent.)
- Smoothness: how promptly can a cyclical turn in the series be distinguished from directional change associated with shorter (mainly irregular) movements? (Score 13.3 per cent.)
- Currency or timeliness: how promptly available are the statistics and how frequently are they reported (monthly, quarterly)? (Score 10 per cent; total 100 per cent, with overall rounding.)

Recalling Fiedler's (1990: p. 137) earlier observation, much statistical work needs to be undertaken to assess and select the individual indicators before compiling the US composite leading economic indicator. Decisions are required regarding the choice of reference cycle with which indicators are compared, and reverse trend adjustment needs to be applied (Shiskin, 1967). Revisions are required from time to time as the structure of the economy changes, especially given the growth in the services sector (Shiskin and Moore, 1967; Auerbach, 1982; Neftci, 1984; Diebold and Rudebusch, 1989; Popkin, 1990; Lahiri and Moore, 1991). The course of the US Composite Leading Index is shown in Figure 7.1. The list below provides the components of this index, as of October 2007:

1 Average weekly hours, manufacturing.
2 Average weekly initial claims for unemployment insurance.
3 Manufacturers' new orders, consumer goods and materials.
4 Vendor performance, slower deliveries diffusion index.
5 Manufacturers' new orders, non-defence capital goods.
6 Building permits, new private housing units.
7 Stock prices, 500 common stocks.
8 Money supply, M2.
9 Interest rate spread, 10-year US Treasury bonds less federal funds.
10 Index of consumer expectations.
                    (Source: The Conference Board, Inc., New York, USA)

In 1995, the Conference Board assumed responsibility for computing the composite indices from the US Department of Commerce. It currently also produces business cycle indices for Australia, France, Germany, Korea, Japan, Mexico, Spain and the UK. In South Africa, a composite leading business cycle indicator is compiled and published by the South African Reserve Bank (Van der Walt, 1982; Venter and Pretorius, 2004; Venter, 2007). As of October 2007, this index is based on 12 components (previously 13) that cover various economic demand and supply variables. The methodology used in constructing the

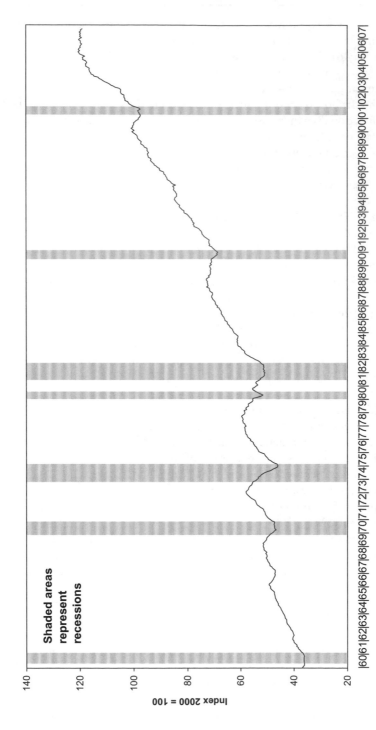

Figure 7.1 US leading economic (indicator, index 2000 = 100).

Source: Conference Board; SA Reserve Bank; MFA Database.

index is based on the Burns-Mitchell model, adapted for local conditions. The course of this leading business cycle index is shown in Figure 7.2. By comparing Figure 7.1 with Figure 7.2, one can observe that economic recessions in South Africa last longer than those in the US.

The Organisation for Economic Co-operation and Development (OECD) was established in 1961, is based in Paris, France, and has 30 member countries. The OECD monitors economic trends, analyses and forecasts economic developments. This body compiles leading economic indices for 29 OECD member countries and for six non-member economies. The OECD integrates business survey data with quantitative indicators in compiling Composite Leading Indicators (CLI data). It employs a modified version of the methodology pioneered by Burns and Mitchell (OECD, 2007). There are several other research bodies one could mention in this context, e.g. the Economic Cycle Research Institute (ECRI) and the Foundation for International Business and Economic Research (FIBER), both based in New York. Researchers in several countries have devised quantitative leading indicators for the construction sector (Van Miltenburg and Romijn, 1993; Snyman, 1994). Akintoye, Bowen and Hardcastle (1998) examined macroeconomic leading indicators of UK construction contract prices. They found that the unemployment level, construction output, industrial production and the ratio of price to cost indices in manufacturing are consistent leading indicators of construction prices.

Summary: Quantitative leading economic indicators

The cyclical indicator approach is neatly summarized by the Conference Board (2007: p. 2). The composite indices are the key elements in an analytical system designed to signal peaks and troughs in the business cycle. The leading, coincident and lagging indices are essentially composite averages of between four and twelve individual leading, coincident, or lagging indicators. They are constructed to summarize and reveal common turning point patterns in economic data in a clearer and more convincing manner than any individual component, primarily because they smooth out some of the volatility of individual components. Historically, the cyclical turning points in the leading index have occurred before those in aggregate economic activity, while the cyclical turning points in the coincident index have occurred at about the same time as those in aggregate economic activity. The cyclical turning points in the lagging index generally have occurred after those in economic activity. Empirical research has shown that the ratio of the coincident index to the lagging index tends to have long leads in the business cycle.

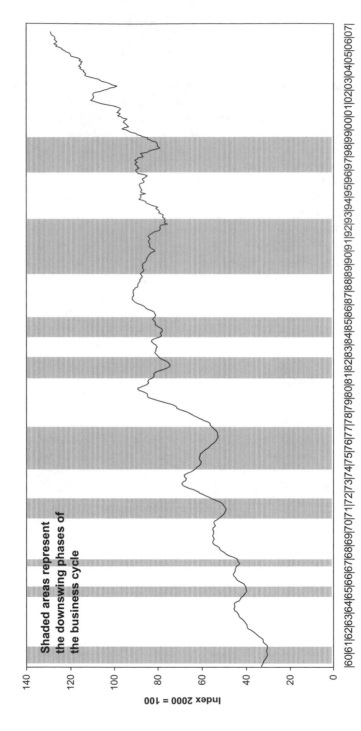

Figure 7.2 South African Reserve Bank composite leading indicator (index 2000 = 100).

Source: SA Reserve Bank; MFA Database.

Qualitative leading economic indicators

Origins of qualitative leading economic indicators

In contrast to quantitative data (e.g. industrial production), that are measured on an interval or ratio scale, qualitative data are measured on a nominal or ordinal scale. Subjective evaluations and judgements made by individuals of variables such as business confidence are expressions of qualitative information. Opinions are gathered by means of surveys; hence the approach is often referred to as the 'business survey method'. Questions can be framed to elicit a response 'increase, no change, decrease' in a variable or an evaluation 'good, satisfactory, bad' or 'favourable/unfavourable'. It is important to note that questionnaires are not structured to gather information on 'how much', but 'whether'. Piatier, a leading French researcher, referred to this survey technique as 'statistique sans chiffres' (statistics without figures). Strigel (1981: p. 25) explained this apparent contradiction by observing that '. . . the determination of that frequency distribution of positive and negative evaluations of qualitative anticipations is also a statistical method'. He added (1981: p. 25) that these data can be '. . . aggregated and converted into time-series so that they can be compared with respective time-series of quantitative statistics'. As these variables reflect the opinions of the survey respondents, they are often referred to as 'soft data', in contrast to quantitative or 'hard data'.

This survey method is widely known as the 'Konjunkturtest'. Strigel (1981: p. 9) mentions that statistics of a qualitative type have been in existence in German agriculture for more than 100 years. In 1927 a start was made in the US with surveys of a qualitative nature by the Regional Shippers Advisory Boards of the Association of American Railroads. Hupkes (1961: p. 3) noted that already in 1928 Wageman had foreseen the possibility of continuously collecting data directly from individual firms to be used for the purposes of business forecasting. Nevertheless, Haberler (in Hupkes, 1961: p. 13) thought this to be a 'very questionable procedure'. According to Strigel (1981: p. 11), it was only after World War II that trade cycle surveys came into their own. In 1948, Katona began the first consumer survey in the US. At the beginning of the 1950s, new activity in this field of economic research was recorded in Europe, particularly in the then Federal Republic of Germany, in France and in Italy. Piatier (1979: p. 1) observed that the first three surveys were undertaken in 1949 by Dr Karl Wagner in Munich, by Prof G. Tagliacarne in Rome and himself in Paris. They worked independently and met in 1950 to find that their pre-occupation with survey data was remarkably similar, viz. as an aid to short-term forecasting.

The interest in this field was such that these economists, representing three European institutions (Ifo (Institute for Economic Research), Munich; Italian Chambers of Commerce, Rome; National Institute for Statistics, Paris), met in Paris in 1952 and founded an international committee for studies on business survey methods, called CIMCO. In 1960 the name of this organization was changed to 'Centre for International Research on Economic Tendency Surveys' (CIRET). For several years (1960–1966) there was a small research centre under the direction of Prof H. Theil at the Econometric Institute of the Netherlands School of Economics in Rotterdam. Prof O. Anderson then continued this work at the University of Mannheim and later at the University of Munich, Germany. In 1971, these activities were merged with the CIRET Information and Documentation Centre at the Ifo Institute for Economic Research, Munich. CIRET was affiliated to the Ifo Institute through a small secretariat and several years ago moved to Zurich in Switzerland where it currently operates under the stewardship of the Eidgenössische Technische Hochscule (ETH) at the KOF Swiss Economic Institute. CIRET has more than 500 members from more than 50 countries and organizes forecasting conferences biennially.

In the US, the well known University of Michigan's Survey Research Center, Ann Arbor, undertakes consumer sentiment surveys. The pioneering work was undertaken by George Katona in the 1940s and early 1950s. According to Curtin (2002: p. 2), Katona developed confidence measures as a means to 'directly incorporate empirical measures of income expectations into models of spending and saving behaviour'. Katona distinguished between consumers' (financial) ability to buy and their willingness to buy. He hypothesized that spending would increase when people became optimistic, and precautionary saving would rise when they became pessimistic (Curtin, 2002: p. 2). In a worldwide review, Curtin (2004: p. 2) observed that consumer spending accounts for one-half to two-thirds of all spending in market-based economies, so even small changes in the expenditures of households can have an impact on the economy.

Benefits of business surveys

Over the years, business surveys have gained acceptance and respectability as a means of business cycle analysis. This could be ascribed, in large part, to the unique benefits of business surveys. Strigel (1981, 1990) mentions several of these benefits:

- Broader information: additional information is gathered to supplement the available database of quantitative statistics.
- Efficiency: questionnaires are fairly simple to complete and can be returned with a minimum loss of time. This could be especially beneficial when forecasting turning points in the business cycle.
- Fuller picture of trends: a synchronization of data becomes possible between (say) the production and distribution sectors in the economy.
- Awareness of limitations: the fact that data collection and data analysis are undertaken by the same institution has certain benefits because researchers are constantly aware of the limitations of the survey results.

- Faster adaptation: the survey method is flexible and questionnaires can readily be adapted to ascertain the effects of economic policy changes.
- Greater trust develops between surveying organization and individual firm. Frank answers can be provided in such an 'information community'.

By stressing the benefits of business surveys, one should not gain the impression that qualitative data are 'better' than quantitative data. However, it is true that business survey data can be compiled and released before most quantitative economic indicators become available. An added benefit is that the survey data are final, whereas quantitative data are normally subject to revision. Though it is true that business survey data do impart knowledge about trends, they do not reflect levels *per se*. Better then to use both qualitative and quantitative data in conjunction with one another and integrate their findings with the objective of improving business cycle analysis.

Development stages of qualitative business surveys

It was with this goal in mind that business surveys underwent considerable development in the post-war period. Several main streams in the theoretical and empirical development of qualitative business surveys, given below, can be identified (Aiginger, 1977; Strigel, 1981).

- Descriptive: research papers set out to justify the use of qualitative surveys in the eyes of a sceptical audience.
- Quantification: research focused on the quantification of qualitative survey data and the related problem of the indifference interval in surveys (i.e. what percentage of respondents report 'no change?').
- Accuracy testing: empirical investigations of the variables were undertaken, mainly with respect to consistency and rationality.
- Business cycle analysis: researchers determined the prognostic applicability of survey indicators.
- Econometric modelling: with a sufficiently long database, survey indicators were incorporated into econometric models and into systems of cyclical indicators.
- Behavioural studies: the survey data are employed to carry out extensive studies of the behaviour of business people and consumers.
- Early warning system: researchers devised an integrated qualitative and quantitative cyclical analysis system.
- International co-ordination: international experience is evaluated to arrive at a conclusive opinion on the optimum formulation of survey questions and the best possibilities for using trade cycle (i.e. business) surveys.

Initially, research focused on business conditions in the manufacturing industry, on price expectations, on consumer sentiment and on surveys of investment intentions. Surveys of innovation activities and new technologies were developed in the early 1990s and research findings were presented at CIRET conferences.

Summary: Qualitative leading indicators

Qualitative business cycle indicators are subjective assessments by business people or consumers and are designed to reflect trends, but not levels, of economic activity. Typical variables include evaluations of business confidence or changes in business conditions, production output, numbers of hours worked, inventories, availability of labour and materials, employment and profitability. Consumer sentiment surveys monitor consumer confidence, the improvement or the deterioration in the financial position of consumers, consumers' expectations 6 months hence, and a host of other useful survey responses. Business and consumer surveys have the advantage of being available before the majority of quantitative indicators. Therefore, they can be integrated purposefully with more traditional quantitative cyclical indicators in short-term forecasting models. In South Africa, Smit (1982) evaluated the usefulness of opinion survey data in econometric models.

Leading economic indicators in practice

Fiedler (1990: p. 137) joined the Conference Board in 1975, and at time of writing was the Vice President and Economic Counsellor. He stated that he had never known any able forecasters who did not use the quantitative cyclical indicators as part of their methodology. Nor had he ever known any good students of the indicators who did not also use other information as part of their forecasting process. He asserts that the leading economic indicators are well regarded and widely known. From the 1960s onward, the US leading, coincident and lagging indicators were published on a monthly basis in the *Business Conditions Digest*, under the auspices of the Department of Commerce. In 1995, the Conference Board assumed responsibility for compiling the composite indices. Given Internet access in modern times, the dissemination of these useful indicators to professional practitioners has been enhanced. The indices are also widely quoted in the media. For example, the US leading index appears regularly in the statistical analysis section of *The Economist*, London, making it readily available to a wider readership.

According to Moore (1983) and Fiedler (1990: p. 137), many forecasters dismiss lagging indicators as being inconsequential. Yet, structural imbalances and distortions develop during the late stages of a cyclical expansion and help bring on the first stage of the economic contraction. Lagging indicators, like some of the leading indicators, are measures of the structural imbalances within the business cycle. For example, inventories rising persistently faster than sales warn of a dangerous over-supply of stocks on sellers' shelves. Rising interest rates suggest a developing credit squeeze. Rising labour costs signal a squeeze on profits. All three of these developments are typical of the ingredients from which recessions are made. Thus, the Conference Board monitors the ratio of the coincident index to the lagging index in its monthly bulletins, as this ratio tends to have long leads in the business cycle.

The coincident indicator is designed in such a way that turning points in the coincident index occur at about the same time as turning points in the business cycle. Normally, economic forecasters have a strictly professional agenda when studying the movements in the coincident indicator. However, sometimes it is used for purposes of a different kind, for example, to predict whether the sitting president of the US would be re-elected, as in the 'Yale forecasting model'. According to Zarnowitz (2004), a doctoral candidate at Yale University studied the interrelationship between business cycles and political events in the US and made an interesting finding. He established that if the US Index of Coincident Economic Indicators was at a higher level at the time of the re-election than the level of the indicator at the point when the incumbent was first elected, then the incumbent was subsequently re-elected. If the index was lower, then the sitting president lost the election for his second term in office. This has been the case, *without exception*, during the past 64 years. Of course, this is a reflection of the fact that a growing economy is beneficial to a sitting president's chances of being re-elected, and probably explains President Bill Clinton's mantra 'It's the economy, stupid'.

Turning now to qualitative cyclical indicators, it is noteworthy that consumer survey data feature prominently in several composite leading indices. In the case of the US Leading Economic Indicator, the Michigan Index of Consumer Expectations features as one of the 10 component indices. The OECD incorporates a consumer confidence indicator in their model. In South Africa, the SA Reserve Bank incorporates three qualitative variables derived from business surveys in their model, consisting of 12 component time series. These three subcomponents relate to business confidence, the average number of hours worked and the volume of orders received (Venter, 2007).

An endogenous test is undertaken when qualitative time series are compared to one another, as, for example, when survey responses reflecting *current* business conditions are compared to those reflecting *expectations* of business conditions. In cases where qualitative time series are compared to their respective counterparts measuring quantitative data, it is known as an exogenous test. In simple terms, soft data are being compared to hard data. Given that the survey results are, in most cases, available before the publication of quantitative data, this implies that forecasters have early warning signals regarding future trends in the hard data (Snyman and Martin, 2006; CESifo World Economic Survey, 2007). A good example of such a comparison is given in Figure 7.3. The Ifo Economic Climate for the euro area reflects data for 13 member countries and is based on the responses of almost 300 professional forecasters. As a rule, the trend in the Ifo Economic Climate indicator correlates well with the actual business-cycle trend for the euro area, measured in annual growth rates of real

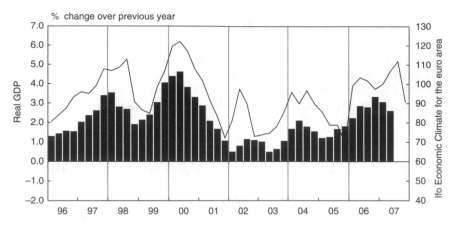

Figure 7.3 Economic growth and Ifo Economic Climate for the euro area (index 1995 = 100) (arithmetic mean of judgement of present and expected economic situation).

Sources: Eurostat, Ifo World Economic Survey (WES) III/2007.

gross domestic product. The recent drop in the Ifo indicator suggests slower growth in the euro area.

In South Africa, Snyman (1994, 1995) devised a leading indicator for the South African building industry by applying the NBER scoring methodology to business survey data compiled by the Bureau for Economic Research, Stellenbosch University. Thirty-three cyclical indicators from a population of 58 indicators were found to be suitable for inclusion in this composite indicator. The effective median lead times of turning points in the Composite Leading Indicator for the South African Building Industry (CLIBI) are given below. Turning points were compared to the construction sector reference cycle compiled by the South African Bank.

- Median lead (−) at troughs   −2.5 months.
- Median lead (−) at peaks   −8.5 months.
- Median lead (−) at all turns   −7.0 months.

These findings support Strigel's (1981: p. 20) assertion that '. . . the lead times of leading indicators usually are substantially longer at the upper turning points than at the lower turning points'. In updated form, this cyclical indicator is shown in Figure 7.4. There is a fairly close correspondence between cyclical swings in the CLIBI and the (smoothed) annual percentage changes in the level of total investment in buildings, as compiled by the SA Reserve Bank. This relationship, depicted in Figure 7.5, is another example of an exogenous test (Snyman, 1999). Both these indicators reflect the current resurgence in the fortunes of the South African building industry.

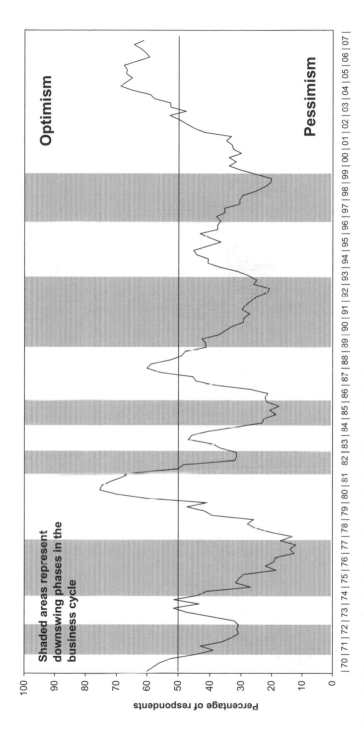

Figure 7.4 Composite leading indicator for the South African building industry.

Source: Bureau for Economic Research; SA Reserve Bank; MFA Database.

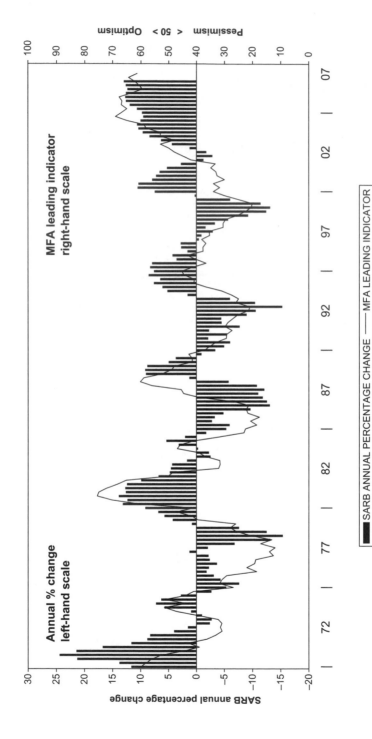

Figure 7.5 MFA leading indicator compared to annual percentage change in total investment in buildings.

Source: FNB/BER; SARB; MFA Database.

Building cycles

Building cycles, by their very nature, are longer than business cycles. Hillebrandt (1974: p. 21) quotes J. Parry Lewis' (1965) painstaking analysis of building cycles in Britain from 1700, who found that there are distinct long-term fluctuations in construction activity, often regional in character. In providing an explanation for these phenomena, he found the '. . . key . . . in population, credit and shocks . . .'. The shocks he had in mind were, for example, bad harvests or war, which will have repercussions on the natural increase in population and on migration.

Migration features prominently in the analysis of Simon Kuznets (1930, 1958), a Nobel Prize Winner for Economics awarded in 1971. According to Abramovitz (1966: p. 521), Kuznets showed that, in the case of the US, '. . . production and price series, with primary trends eliminated and the influence of business cycles attenuated or smoothed by moving averages, exhibited pronounced wave-like undulations'. To recognize him for his long record of work, W. Arthur Lewis (Nobel Prize Winner for Economics, 1979) suggested that these 15–25 general waves in economic life be referred to as *Kuznets cycles* (Abramovitz, 1966: p. 520). These *cycles* or *waves* were clearly identified in 'population-sensitive' capital formation, notably railroad investment and urban residential building.

In detailed research, Kuznets showed that waves in immigration and other demographic factors, such as household formation, as well as changes in age composition, marriage and birth rates, gave rise to long swings in the rate of economic growth. These factors could be regarded as the primary cause of fluctuations in transport and housing investment. He also indicated that a tight labour market stimulates immigration and otherwise encourages household formation and, therefore, the demand for residential building. Abramovitz (1968) has suggested that, with the efflux of time, the Kuznets cycle in the US has passed away. Kuznets, himself, passed on in 1985. One wonders whether, given an influx of Mexicans to the US in recent times, there would be evidence of the development of a new Kuznets wave.

In South Africa, Kilian and Snyman (1984) found evidence to show that Kuznets cycles exist in that sector of the building market where one would expect to find them, i.e. in private housing. First, the long-term growth rate in investment in private residential building was established. Second, the percentage deviation from this growth trend was calculated. At that point, the results of this investigation showed that there was a strong positive correlation between fluctuations in migration patterns and investment in housing. In updated form, these Kuznets cycles can be viewed in Figure 7.6. Two cycles of 16 years and one cycle of 20 years are evident in post-war South African data. Also, observe that each trough in the series occurs at a higher level than the previous trough. No doubt, this can be ascribed to rapid population growth during this period. Finally, the data suggest the emergence of a fourth Kuznets cycle in recent times.

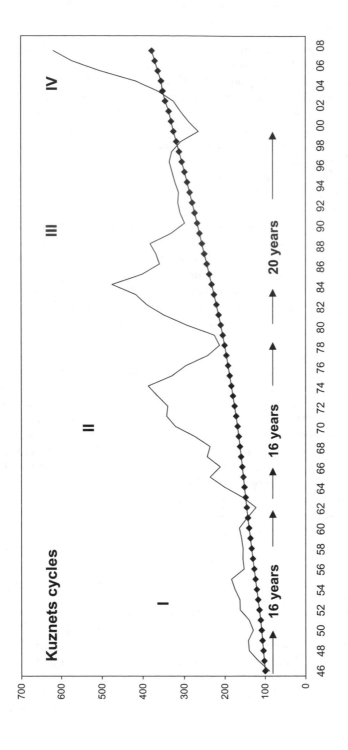

Figure 7.6 Kuznets cycles in South Africa private housing investment vs. population (index 1946 = 100).

Source: SARB, StratsSA; MFA Database.

The findings of Kilian and Snyman (1984) were corroborated by Rode (1984) who found a similar 16 year property cycle during the period 1962–1978 in South African data of real Johannesburg office rentals.

Regarding their amplitude, it is appropriate to mention that business cycles and building cycles interact in such a way that they support or retard each other over the course of the cycle. As Hindle (1996: p. 427) noted, Cloete (1990) asserted that the business cycle can be affected by the longer building cycle because of the importance and size of the industry to the national economy. Cloete (1990: p. 14) avers:

> In the South African economy, as in other economies, statistics show that business cycle upswings which coincide with an expansion in building activity tend to be relatively strong and prolonged, while those occurring during periods when building activity is declining tend to be relatively short and anaemic. Similarly, business cycle downswings occurring when building activity is contracting tend to be relatively severe and prolonged, while those coinciding with the expansion phases of the building cycle tend to be relatively short and mild.

These arguments support the widely held view that the amplitude of fluctuations in construction exceeds fluctuations in the manufacturing and services sectors of the economy (Kilian, 1976; Rosen, 1979; Bon 1989; Snyman, 1994; Mills, 1995; Akintoye, 1996; Hindle, 1996). According to the World Bank (1984), this tendency is inherent in the demand structure of capital goods industries where relatively small changes in demand by consumers will cause the production capacity to be expanded or contracted at a considerably higher rate. The World Bank asserts that fluctuations in construction tend to be greater in developing than in developed countries.

Kilian (1976) and Hindle (1993, 1996) are in agreement that there is not just 'one' building cycle. Killingsworth (1990: p. 11), Akintoye and Skitmore (1994: p. 11), and Hindle (1996: p. 432) suggest that research should be directed to identify the demand cycle for the various sub-sectors or different markets for construction. Their suggestion implies a search for leading indicators that will facilitate prediction of the turning points in the demand cycle for residential, industrial and commercial building. This is important because fluctuations in the level of construction directly affect changes in cost inputs and tender prices. Thus, improved forecasting methods of cyclical changes in demand levels could enhance cost planning of construction projects (Snyman, 1980, 1989, 2007; Bowen and Edwards, 1985; Taylor and Bowen, 1987; Runeson, 1988; Hindle, 1993; Mills, 1995). So one ends where one started, by observing that the relationship between demand and prices is a recurring theme in the literature on the built environment.

Summary

In this contribution, the author has tried to show that research surrounding leading economic indicators has a rich history. It is intertwined with research into business cycles. In the US, work started in the early 1900s and followed a quantitative approach. In Europe, during the post-war period, the emphasis fell on qualitative data, gathered by means of business surveys. Essentially, these indicators were compiled as an aid to short-term economic forecasting. As most forecasters are aware, establishing one's current position in the business cycle is not that straightforward. Therefore, better results seem to be achieved when combining quantitative and qualitative data in economic forecasting models. This is done formally in econometric models, or informally in time series analysis. Today, leading indicators are firmly grounded in theory, and they are well regarded and widely used. Yet, they are not infallible, and false signals do occur. Hence, as Taleb (2007) has shown, forecasters need to be wary of 'black swans' appearing on the horizon. Building professionals need to take special cognisance of economic shocks, such as a sudden rise in interest rates or tighter credit conditions, because they are operating in a capital goods industry where cycles are longer and more pronounced than in other sectors of the economy. For the entrepreneur, there are opportunities aplenty; for the researcher in the built environment, there are many unanswered questions; for the practising construction economist, much research is still required concerning the interaction between business cycles and building cycles in developed and developing countries and how these affect the formation of costs and prices.

References

Abramovitz, M. (1961) The nature and significance of Kuznets cycles. *Economic development and Cultural Change No 9*. Reprinted in Gordon, R.A. and Klein, L.R. (eds.) (1966) *Readings in Business cycles selected by a committee of the American Economic Association*. London: George Allen & Unwin Ltd.

Abramovitz, M. (1968) The passing of the Kuznets cycle. *Economica*, August, 349–367.

Aiginger, K. (1977) *The use of survey data for the analysis of business cycles*. CIRET Studies No 24. CIRET: Munich.

Akintoye, A. (1996) Volatility and cycles in commercial construction market. *Economic Management of Innovation, Productivity and Quality in Construction Vol 1 CIB W55 Building Economics 7th International Symposium*. Zagreb, Croatia.

Akintoye, A., Bowen, P.A. and Hardcastle, C. (1998) Macro-economic leading indicators of construction contract prices. *Construction Management and Economics*, 16, 159–175.

Akintoye, A. and Skitmore, M. (1994) Models of UK private sector quarterly construction demand. *Construction Management and Economics*, 12, 3–13.

Auerbach, A.J. (1982) The index of leading indicators: 'Measurement without theory', thirty-five years later. *Review of Economics and Statistics*, 64, 589–595.

Boehm, E.A. (1990) Understanding business cycles today: a critical review of theory and fact. In: Klein, P.A. (ed) *Analyzing Modern Business Cycles: Essays Honoring Geoffrey H Moore*. M E Sharpe, Inc: New York.

Bon, R. (1989) *Building as an Economic Process: an Introduction to Building Economics*. Prentice-Hall, Inc.: Englewood Cliffs, New Jersey.

Bowen, P.A. and Edwards, P.J. (1985) Cost modelling and price forecasting: practice and theory in perspective. *Construction Management and Economics*, 3, 199–215.

Burns, A.F. and Mitchell, W.C. (1946) *Measuring Business Cycles*. National Bureau of Economic Research: New York.

CESifo (2007) *World Economic Survey 2007 6(3)*. Ifo institute for Economic Research: Munich.

Cloete, J. (1990) *The Business Cycle and the Long Wave with Special Reference to South Africa*. The Institute of Bankers in South Africa: Johannesburg.

Conference Board (2007) *US Leading economic indicators and related composite indexes for October 2007*. (http://www.conference-board.org/economics/bci Website accessed 27 November 2007).

Curtin, R.T. (2002) Consumer confidence in the 21st Century: changing sources of economic uncertainty. *Paper read at the 26th CIRET Conference*. Taipei. October.

Curtin, R.T. (2004) Consumer sentiment surveys: worldwide review and assessment. *Paper read at the 27th CIRET Conference*. Warsaw. September.

Diebold, F.X. and Rudebusch, G.D. (1989) Scoring the leading indicators. *Journal of Business*, 62(3), 369–391.

Fiedler, E.R. (1990) The future lies ahead. In: Klein, P.A. (ed) *Analyzing Modern Business Cycles: Essays Honoring Geoffrey H Moore*. M E Sharpe, Inc: New York.

Greenspan, A. (2007) *The Age of Turbulence. Adventures in a New World*. Allen Lane (Penguin): London.

Hillebrandt, P.M. (1974) *Economic Theory and the Construction Industry*. Macmillan: London.

Hindle, R.D. (1993) The effects of the short-term business cycle on the construction industry. In: Manso, E.D. and Plat, H. (eds) *CIB W55 Symposium on the Economic Evaluation and the Built Environment. Laboratorio Nacional de Engenharia Civil, Vol 2*. Lisbon. September.

Hindle, R.D. (1996) What are we looking for in the search for building cycles? *Economic Management of Innovation, Productivity and Quality in Construction Vol 1 CIB W55 Building Economics 7th International Symposium*. Zagreb, Croatia.

Hupkes, G.J. (1961) *An Evaluation of the Opinion Survey Method of Business Forecasting as Applied by the Bureau for Economic Research*. Bureau for Economic Research: Stellenbosch University.

Kilian, W.F. (1976) *On Stabilising Construction: an Investigation into the Characteristics and Cyclical Fluctuations of the Construction Industry to Recommend Possible Stabilising Measures*. Bureau for Economic Research: Stellenbosch University.

Kilian, W.F. and Snyman, G.J.J. (1984) The Kuznets and Kondratieff cycles and their relevance in property. *Juta's South African Journal of Property*, October/November.

Killingsworth, Jr. R.A. (1990) A preliminary investigation into formulating a demand forecasting model for industrial construction. *Cost Engineering*, 32(8), 11–15.

Klein, P.A. (1983) The neglected institutionalism of Wesley Clair Mitchell: the theoretical basis for business cycle indicators. *Journal of Economics Issues*, 17(4), 867–899.

Klein, P.A. (ed) (1990) *Analyzing Modern Business Cycles: Essays Honoring Geoffrey H Moore*. M E Sharpe, Inc: New York.

Klein, P.A. and Moore, G.H. (1985) Monitoring growth cycles in market-oriented countries: developing and using international economic indicators. *National Bureau of Economic Research Studies in Business Cycles No 26*. Ballinger: Cambridge, Mass.

Koopmans, T.C. (1947) Measurement without theory. *Review of Economics and Statistics*, 29, 161–172. Reprinted in Gordon, R.A. and Klein, L.R. (eds) (1965) *Readings in Business Cycles*. Illinois: Irwin.

Kuznets, S.S. (1930) *Secular Movements in Production and Prices: Their Nature and Bearing upon Cyclical Fluctuations*. Houghton Mifflin Co: New York.

Kuznets, S.S. (1958) Long swings in the growth of population and in related economic variables. *Proceedings of the American Philosophical Society. Vol CII, No 1*.

Lahiri, K. and Moore, G.H. (1991) *Leading Economic Indicators: New Approaches and Forecasting Records*. Cambridge University Press.

Lewis, J.P. (1965) *Building Cycles and Britain's Growth*. Macmillan: London.

Mankiw, N.G. (1990) A quick refresher course in macroeconomics. *Journal of Economic Literature*, 28.

Mills, A. (1995) A history of building industry trade cycles. *Paper presented in May at the AIQS/ICEC symposium at Broadbeach, Australia*. Reprinted in *The Building Economist*, September, 1645–1660.

Mintz, I. (1974) *Dating United States Growth Cycles*. Explorations in economic research. NBER I (Summer).

Moore, G.H. (1983) *Business Cycles, Inflation and Forecasting*, 2nd edition. Ballinger: Cambridge, Massachusetts.

Moore, G.H. and Cullity, J.P. (1993) The historical record of leading indicators – an answer to 'measurement without theory'. *Paper read at the 21ˢᵗ CIRET Conference*. Somerset West, South Africa. October.

Neftci, S. (1984) Are economic time series asymmetric over the business cycle? *Journal of Political Economy*, 92, 307–328.

Oppenländer, K.H. (1993) Narrowing the induction gap: measurement without theory? *Paper read at the 21ˢᵗ CIRET Conference*. Somerset West, South Africa. October.

Organisation for Economic Co-operation and Development (2007) OECD Statistics: leading indicators. (http://www.oecd.org – accessed 27 November 2007).

Piatier, A. (1979) Business cycle surveys – their utilization for forecasting. *Paper read at the 14ᵗʰ CIRET Conference*. Lisbon. September.

Popkin, J. (1990) Why some of the leading indicators lead. In: Klein, P.A. (ed) *Analyzing Modern Business Cycles: Essays Honoring Geoffrey H Moore*. M E Sharpe, Inc: New York.

Rode, E.G. (1984) *Annual Rental Review of Old Mutual Properties*. Old Mutual Ltd: Cape Town.

Rosen, K.T. (1979) Cyclical fluctuations in residential construction and financing. In: Lange, J.E. and Mills, D.Q. (eds) *The Construction Industry: Balance Wheel of the Economy*. Lexington Books: Lexington, Massachusetts.

Runeson, G. (1988) Methodology and method for price-level forecasting. *Construction Management and Economics*, 6, 45–55.

Shiskin, J. (1967) Reverse trend adjustment of leading indicator. *Review of Economics and Statistics*, 49, 45–49.

Shiskin, J. and Moore, G.H. (1967) *Indicators of Business Expansions and Contractions*. National Bureau of Economic Research: New York.

Smit, E. van der M. (1982) *Die waarde van opnamedata in ekonometriese modelle met spesifieke verwysing na die opiniatiewe data van die Buro vir Ekonomiese Ondersoek. (The value of opinion survey data in econometric models with specific reference to the opinion survey data of the Bureau for Economic Research).* Unpublished DComm dissertation, Stellenbosch University.

Snyman, G.J.J. (1980) On building cost indicators. *BER Building Survey No 44.* Bureau for Economic Research: Stellenbosch University.

Snyman, G.J.J. (1989) How the business cycle influences building costs. *Juta's South African Journal of Property*, 5(3).

Snyman, G.J.J. (1994) *The development of leading indicators for the South African building industry using qualitative and quantitative data.* Unpublished PhD dissertation, Department of Construction Economics and Management, University of Cape Town.

Snyman, G.J.J. (1995) New leading indicators for the South African building industry. In: Oppenländer, K.H. and Poser, G. (eds) *Business cycle surveys: forecasting issues and methodological aspects. Selected papers presented at the 22nd CIRET conference, Singapore, 1995.* Avebury: Aldershot.

Snyman, G.J.J. (1999) Leading indicators for the South African building industry: the effect of the Reconstruction and Development Programme. In: Oppenländer, K.H., Poser, G. and Schips, B. (eds) *Use of survey data for industry, research and economic policy. Selected papers presented at the 24th CIRET conference, Wellington, New Zealand.* Ashgate: Aldershot.

Snyman, G.J.J. (2007) *Using knowledge of the business cycle to forecast building costs. Paper read at the Construction Management and Economics 25 Conference.* University of Reading, UK.

Snyman, G.J.J. and Martin, C.H. (2006) Consumer confidence in South Africa: an exogenous test. Comparison with house prices, consumer spending and homebuilding. *Paper read at the 28th CIRET Conference.* Rome, Italy. September.

Strigel, W.H. (1981) *Essays on trade cycle surveys.* CIRET studies No 30. CIRET: Munich.

Strigel, W.H. (1990) Business cycle surveys: A new quality in economic studies. In: Klein, P.A. (ed) *Analyzing Modern Business Cycles: Essays Honoring Geoffrey H Moore.* M E Sharpe, Inc: New York.

Taleb, N.N. (2007) *The Black Swan: the Impact of the Highly Improbable.* Random House: New York.

Taylor, R.G. and Bowen, P.A. (1987) Building price level forecasting: an examination of techniques and applications. *Construction Management and Economics*, 5, 21–44.

Van der Walt, B.E. (1982) *The identification and evaluation of economic indicators for a study of business cycles in South Africa.* Unpublished DComm dissertation, Faculty of Economic and Political Sciences, University of Pretoria.

Van Miltenburg, H. and Romijn, G. (1993) A composite leading indicator for the Dutch construction sector. *Paper read at the International Symposium on Economic Evaluation and the Built Environment.* Lisbon. September.

Venter, J.C. (2007) Revisions to the composite leading and coincident business cycle indicators. *Quarterly Bulletin, No 244, June 2007.* South African Reserve Bank: Pretoria.

Venter, J.C. and Pretorius, W.S. (2004) Note on the revision of composite leading and coincident business cycle indicators. *Quarterly Bulletin, No 231, March 2004.* South African Reserve Bank: Pretoria.

Vining, R. (1949) Koopmans on the choice of variables to be studied and of methods of measurement. *Review of Economics and Statistics, 31.* Reprinted in Gordon, R.A. and Klein, L.R. (eds) (1965) *Readings in business cycles.* Irwin: Illinois.

World Bank (1984) *The construction industry: issues and strategies in developing countries.* World Bank: Washington, D.C.

Zarnowitz, V. (1985) Recent work on business cycles in historical perspective: a review of theories and evidence. *Journal of Economic Literature*, 23, 523–580.

Zarnowitz, V. (1992) *Business Cycles: Theory, History, Indicators and Forecasting.* University of Chicago Press Ltd: London.

Zarnowitz, V. (1999) Theory and history behind business cycles: are the 1990s the onset of a golden age? *Journal of Economic Perspectives*, 13(2), 69–90.

Zarnowitz, V. (2004) Growth, business cycles and the indicators: a global and historical perspective. *Paper presented at the 27th CIRET conference Warsaw 2004.* (Personal communication).

Zarnowitz, V. and Boschan, C. (1975a) Cyclical indicators: an evaluation of new leading indexes. *Business Conditions Digest.* Bureau of Economic Analysis, US Department of Commerce: Washington, D.C.

Zarnowitz, V. and Boschan, C. (1975b) New composite indexes of coincident and lagging indicators. *Business Conditions Digest.* Bureau of Economic Analysis, US Department of Commerce: Washington, D.C.

Chapter 8

Construction markets in a changing world economy

Stephen Gruneberg

Introduction and overview

Construction markets respond to changes in demand. The various factors that influence the demand for construction in different parts of the world explain the diversity of performance of different construction industries. These underlying conditions combine to influence national income, which, with expectations about future national income, is the main influence determining effective demand for construction services.

Following this brief introduction and overview of the chapter, we discuss these economic and social changes. As these factors influence national income, national income data is used to predict developments taking place in construction markets. However, the data needs to be treated with some caution.

In an international report on the construction industries of several countries, Carassus (2004) argues in favour of a sectoral approach, which takes into account the fact that construction involves a wider number of participants than only contractors and their sub-contractors on site. These participants in the construction process include architects and engineers, facility owners and managers, public authorities, manufacturing suppliers and research and education institutions. This sectoral approach, in principle, would be more useful in assessing the actual contribution of construction activity to gross domestic product (GDP).[1] However, national income statistics and industry data tend to understate the importance of construction as the data is based on an industrial classification consisting only of firms and labour working directly on site.

In this chapter, we compare a large number of countries using constant dollars and exchange rates. This can lead to misleading conclusions. For example, we shall see that the construction market in India is approximately the same size as the Australian construction market, namely $100 billion. This does not take into account that $1 million spent in India would purchase more construction than the same investment in Australia. This is important to bear in mind throughout this chapter, as differences between the poorest and wealthiest nations will tend to exaggerate the real differences. A better international comparison might be based on purchasing power parities in the construction sector but construction data in this form is not available. The data used here is, therefore, of little use in terms of measuring real construction output in the different countries, but is still of use to international contractors or international investors and developers looking to invest. Nevertheless, the figures are indicative and do point out major differences between countries and do account for changes in construction markets taking place throughout the world.

Although economic and social changes are currently raising incomes in many poorer parts of the world, many countries in those regions remain (with several notable exceptions) considerably less affluent than the countries of Western Europe and North America. Indeed, the data suggests that a polarization is continuing to take place, whereby rich countries tend to become relatively richer and poor countries tend to become relatively poorer, with direct implications for their respective construction industries. Nevertheless, the data indicates that some countries are able to transfer from the poorer nations' category to the richer nations' category. We conclude this chapter with a discussion of possible economic implications of rapid economic growth for the construction industry and a brief survey of a number of major construction markets in different parts of the world.

One might expect that, the more developed and diversified an economy, the less significant would be the construction sector. Construction becomes only one of many activities and as the construction industry tends to have relatively low productivity, it forms a smaller percentage of the whole economy. However, the opposite can also be argued. When economies grow, more funding is available for construction projects and private and public demand for improved housing, non-residential buildings and infrastructure increase. As economic activities become more complex and production processes more specialized, more diverse building types are required and more is expected of buildings in terms of comfort, convenience and appearance. The rate of economic growth, population growth and urbanization all contribute to the growth of construction demand and construction firms respond by increasing their capacity and output.

The relationship between the wealth of a country and its spending on buildings is not a new or recent phenomenon. Ancient civilizations such as the Egyptians, Greeks and Incas all displayed their wealth through their palaces, temples and amphitheatres. Rapid changes taking place in the modern global economy affect the incomes and wealth of individual countries and have a significant influence on construction markets.

Global economic growth is dominated by the transformation taking place in Brazil, Russia, India and China, where construction markets have expanded rapidly to facilitate their expansion especially since 2000. Elsewhere, economic growth has lagged behind, especially in many African countries. Many countries which have been viewed as developing countries in the past have increased their national income per head, although they are very often still far behind the average income per head of the mature and developed states in Northern and

Western Europe and North America. According to United Nation's data, in 2005 the average national income per head in Northern Europe was $40,157 and in North America it was $43,124. In Japan, Singapore and the Republic of Korea the equivalent figures were $34,661, $30,159 and $18,164 respectively. The global changes taking place as a result of rapid economic growth are not uniform.

Demographic, political, environmental, economic and technological changes

We begin by considering some of the demographic, political, environmental, economic and technological changes which are taking place in many countries. The size of national construction markets depends on the size of the populations being served and on their level of income. Changes in construction markets depend on changes in the population and changes in national income.

The rate of population increase places demands on construction industries throughout the world. Table 8.1 shows the rate at which populations grew in 2005 in a number of countries.

Those countries with the fastest growing populations (more than 3% in 2005) appear to be either amongst the very wealthiest or the very poorest nations in the world. In the wealthier countries, migration is partly responsible for increases in population as people from other countries seek employment there. The combination of high national incomes per head and rapid population expansion makes these markets particularly strong construction markets. In the poorest countries of the world, a number of factors contribute to rising populations, including traditional extended families, high fertility rates, large families for security in old age, female education issues, health improvements and poverty. In spite of their rapid population expansion, the poorest countries

Table 8.1 Annual population growth rates in 2005. (Source: Summary of population growth rates, data from the United Nations Statistics Division website.)

Growth rates	Number of countries	Examples in order of declining growth rates
Over 3%	22	Qatar, UAE, Eritrea, Afghanistan, Kuwait, Benin, Uganda, Gambia
2–2.9%	45	Angola, Jordan, Kenya, Nigeria, Saudi Arabia, Iraq, Philippines
1–1.9%	73	Bangladesh, Israel, Malaysia, Pakistan, Ireland, India, Brazil, Vietnam, New Zealand, Canada, US
0–0.9%	66	Mexico, China, Norway, Zimbabwe, France, UK, Sri Lanka, Italy, Cuba, Germany
Less than 0%	22	Poland, Hungary, Russian Federation, Ukraine, Georgia

are not able to translate their requirements for buildings into demand for construction, especially outside their capital cities.

China's population grew by less than 1 per cent in 2005. By having few children, families in China have been able to increase their demand for goods and services and this has had the effect of diversifying their economy. At the same time, the Chinese government allowed Chinese factories to be used as employers of Chinese labour in an export-led economy. Although the population growth rate is not particularly high, the increase in construction activity in China has been one of the fastest in the world (see Table 8.6). This is due to the rapid industrialization taking place and the need for ever more buildings and infrastructure to meet the needs of the growing economy.

Following the relaxation of restrictions preventing the free movement of people, Table 8.1 indicates a decline in the populations of countries in Eastern European and in former members of the Soviet Union. The decline in population would appear to be the result of a significant number of people choosing to emigrate. However, although the populations may be in decline, many of these countries are experiencing rapid economic growth and foreign inward investment. The net effect appears to be that construction industries are growing in those countries as they modernize their economies and rebuild their infrastructure, partly as a result of political initiatives taken by the European Union. Political changes in some countries have thus heralded major growth in construction activity, not only in the recent EU accession states of Eastern European but also in Russia.

Political changes can be either beneficial or detrimental to economic development and construction activity. Construction industries thrive best where the political situation is stable, there are profitable opportunities and the risks are largely of a commercial nature. As a consequence of political unrest, conflict and wars, many countries lag behind in their development due to the reluctance of developers to become involved. Large parts of Africa, most notably the Democratic Republic of the Congo, have remained underdeveloped. This also is the case in Afghanistan and parts of Pakistan.

It is important to emphasize the management of country risk due to political uncertainty. Global investors always have a choice and tend to channel their investments to those countries with the least risk. Where political instability is present, returns on fixed capital investments are vulnerable over the longer term. As buildings require an extended period of future cash flows, these are put at risk by the possibility of regime changes and asset seizures. Moreover, where there is political uncertainty, there are greater opportunities for petty corruption and damage to property, which can threaten the financial viability of investments. Political instability therefore has a deterrent effect on investors, which then lowers demand for construction.

Direct government intervention can mean building work and major infrastructure projects are undertaken by the public sector, especially if private investors are unwilling to accept the risk. Construction markets invariably respond.

This has been the case in, for example, Iraq and the Gulf States and above all in China.

As climate change issues become a more prominent feature in global politics and the effects of global warming are seen throughout the world, damage to the built environment through natural disasters is likely to increase demand for construction services. Indeed this is already occurring. Not only are major population centres threatened with building failures caused by extreme weather conditions, the amount of construction work needed to rebuild after tsunamis, flooding, hurricanes and earthquakes creates an opportunity to improve the quality of the built environment to protect people from any future recurrence of the same weather conditions. Building standards are likely to be affected by the need to construct stronger weather-resistant buildings and infrastructure. This also leads to increased demand for construction. As specific areas tend to be destroyed or damaged by sudden unexpected events, the rebuilding programme needed often overwhelms local building industry capacity. Private insurance becomes prohibitively expensive if large-scale natural disasters cause damage over large areas. Government intervention or international aid is required to finance the rebuilding. Even in wealthier countries, such as the US where hurricane Katrina destroyed much of New Orleans in 2005, public sector intervention was required.

Culminating in the publication of the Stern review on the economics of climate change (2007), the concept of sustainable development in the minds of government and private sector developers has also encouraged changes in demand for construction, especially amongst developed nations. Legislation to reduce carbon emissions means that greater pressures have been placed on contractors to produce environmentally friendly buildings. This process demonstrates how wealthy countries tend to demand higher standards than the less wealthy, and building costs increase as a result. Although still in its infancy, these qualitative specifications appear to be increasing in many countries especially in Europe and North America and are yet another change in the world economy affecting the construction industry and construction markets.

The growth of international trade as a result of the General Agreements on Tariffs and Trade (GATT) and the World Trade Organisation (WTO) has meant that many countries have been able to take advantage of their international comparative advantages. In other words, countries were able to specialize in their economic strengths, even when their output is inferior to or more expensive to produce than that of other countries. In this way international trade increases the size of GDPs including the GDP of poor nations. This has had the effect of stimulating those economies with large labour forces. These countries have then been able to trade their increased manufacturing output to create employment and stimulate their own domestic economies. Although incomes remain low in many of these countries, their rates of growth indicate that standards of living have risen significantly since the 1970s. As incomes grew, demand for improved housing also increased and governments were able to provide more infrastructure.

Changes in trading patterns are shifting the centre of economic gravity from Europe and North America towards Asia. Nevertheless, Europe and North America retain educated, skilled and adaptable populations, which imply that, as economic conditions change, their skills will enable them to change and to take advantage of new employment opportunities. This means that their ability to maintain high value-added industrial manufacturing and service activities are likely to generate increased income levels that create continuing demand for construction. Higher income levels lead to higher demand for repair and maintenance as the standard of maintenance tends to increase with income. Moreover, the increased stock of buildings also implies additional demand for repair and maintenance work in the future.

The exploitation of natural resources including oil and mineral wealth has also funded development in countries such as Saudi Arabia, Angola and Venezuela. Although the wealth generated by these natural endowments does not necessarily filter down to the local populations to any great extent, the need to administer production and liaise with national governments can often be seen taking the form of commercial urban developments in capital or major cities. Examples of this can be seen in Nigeria.

Technological change has also induced economies to invest in building projects, because new and more buildings are required. Innovations in hardware and software in the computer industry, and arising out of its applications in industry and commerce, have, for example, had an impact on construction demand in many countries throughout the world. In the hardware industry the demand for dust-free environments has led to new and higher levels of building performance being achieved. As far as software providers are concerned the building types are not particularly distinct from the point of view of the construction technology used. Nevertheless, the specifications for new office accommodation have undoubtedly shifted as a result of the heating and electrical supply requirements of the new technology. The use of the Internet and cheap telecommunications has given rise to a great number of facilities for types of work not known until near the end of the last century. These changes in the use of information technology have been predominant in those countries providing services such as banking, finance and insurance, but have also affected call centres and distribution channels. The changes have given rise to changes in the built environment and hence increased the size of construction markets, often far from where the goods and services are ordered.

In more general terms, technological change and modern global production techniques mean that production has become a global process from the extraction of raw materials to the manufacture of components in different countries to assembly in another and market and sell throughout the world. This has meant that each stage in the production process is now subject to the law of comparative advantage and is often no longer the finished product (or service).

This has led to international financing of inward investment in countries such as India, China and Brazil, and the growth of manufacturing in Asia in particular, with the growth in their construction markets to accommodate increased industrial and commercial activity.

As the world economy becomes increasingly integrated and interdependent in terms of international trade in goods and services, transport infrastructure expansion becomes necessary and increases demand for civil engineering projects. Moreover, as transport technology improves and travel costs decline, the intra-national and international movement of people, as well as goods, increases.

Intra-national movement of people describes the migration of people within countries, often from rural to urban areas. Countries such as Mexico, Uganda and the Philippines are all examples of rapid urbanization as growing populations migrate towards the towns and cities. This process of rapid urbanization places great strains on urban planners and the construction industry to deliver appropriate infrastructure, housing and places of work.

The international migration of people from the poorer regions of the world in search of employment in the wealthier regions also contributes to increased demand for construction. This international movement of people reflects the movement of goods and services and increases the interconnectivity of the global economy. This has a further impact on the demand for construction as hotels provide for temporary visitors such as tourists and business managers while housing and other facilities are needed for people moving between countries on a more permanent basis.

The movement of people within wealthy regions such as the US or Europe is facilitated (though not necessarily without social and political friction). People within Europe are free to move between member states of the EU for leisure, work or retirement. Similarly, people may move freely with states or migrate between states in the US. There is also a movement of migrants from poorer regions to wealthier countries, such as from South America, Africa and Asia to the US and to Europe.

Migration of people is in itself no different from the movement of goods and services between countries. One purpose served by a market price is to send a signal to attract goods, services and resources to where they can command the highest prices or wages. However, migration has implications for the construction industries in both the host destination and the places of origin.

As populations move from one region to another, they increase the demand for accommodation at the destination point. The increase in demand increases the price of accommodation and this in turn increases house and land prices. It also increases the demand for construction. In the country of origin, the departure of a significant proportion of the local population has the effect of depressing property and land prices in the short term. Apart from political asylum seekers, migrants most usually leave their homeland for economic reasons, namely to achieve higher incomes than those on offer in their own area or region. They often aspire to returning to their country of origin. As a consequence of this desire to return, income earned abroad is often partly repatriated for investment purposes, most usually for a home in their retirement. Thus, a proportion of repatriated income may be funnelled into land, property investment and domestic construction projects, increasing prices of all three over the longer term.

The various social and economic factors discussed above are present to a varying extent in every country. No one factor can be said to be the cause of the economic conditions affecting a nation's construction industry. Nevertheless, these factors taken together contribute in one way or another to the level and distribution of national income in each country and it is the level of national income per person and expectations concerning the growth and sustainability of that income that ultimately drives construction demand and construction markets. This model of the effects of a changing world economy on construction markets in different countries is illustrated in Figure 8.1.

The pattern of change

Population and national income

The model given in Figure 8.1 assumes the level of demand for construction services depends on the value of the national income per head, assuming national income is distributed equally amongst the population. Hence, at its simplest, the greater the GDP per head, the greater will be the demand for

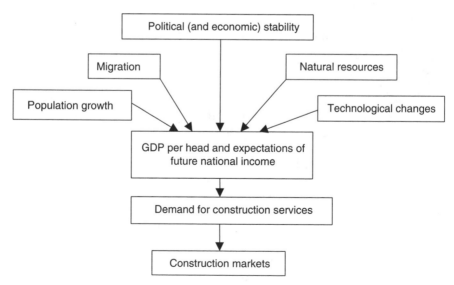

Figure 8.1 The factors influencing construction markets.

construction services. However, the data provides a more complicated picture. Table 8.2 shows the spread of national income per head of all regions and subregions in the world between 2000 and 2006. In 2006, the GDP per head in Asia and Africa was $162 and $1,158 respectively, well below Oceania, Europe and the Americas, all above $19,510. The annual GDP per head in the subregions ranged from as little as $346 in East Africa to as much as $43,124 in Northern America. It should therefore not be surprising that many migrants risk their lives to leave the poorer regions in search of work in the wealthy parts of the world, often as construction labour.

If income per head in 2006 is taken as an indicator of current construction demand, then the largest construction industries can be expected in North America, Northern Europe, Western Europe, and Australia and New Zealand. The worst performing subregions are Eastern Africa, Western Africa and South Central Asia. There are wide variations within regions and as we have noted above, there are important exceptions regarding the size and growth rates of certain countries especially in Asia.

As the data in Table 8.2 is given in current prices, inflation is not taken into account. In order to overcome this problem, Table 8.3 shows the ratio of GDP per head of all regions and subregions between 1970 and 2006 over the weighted world average for each year.

Changes in the world economy since 1970 are shown in Table 8.3. In global terms, there has been a polarization of income per head. The wealthiest regions of the world have increased their relative incomes compared to the poorer regions. Even poorer subregions that have performed well since 1970, such as Eastern Asia, had an income per head that, in 2006, was still only 75 per cent of the world's average. In contrast, in Northern Europe incomes per head increased from 2.87 times the world average to 5.45, catching up with North America. In North America, national income per head was 5.85 times the world average in 2006. Meanwhile the average income per head in Africa declined from 29 per cent of the global average in 1970 to only 16 per cent in 2006.

Within regions, economic performance can vary greatly between countries. China, for example, has only increased its ratio of GDP per head from 13 per cent to 28 per cent of the global average. However, the size of the Chinese economy and direct government intervention in construction projects mean that the construction industry in China has grown far more than the national income figures alone would suggest. According to Crosthwaite and Connaughton (2008), in 2007 China's construction market at over $600 billion was the third largest in the world, behind the US and Japan. One reason for this is that income is unequally distributed and a high proportion of its citizens earn far below the national income per head.

Based on data published by the UN (2008), India's relative position actually declined from only 13 per cent of the average GDP per head in 1970 down to 11 per cent in 2006 in spite of the rapid growth in the economy. In India, the population has grown rapidly and economic growth has not been universal.

Table 8.2 Estimates of GDP per capita in US dollars at current prices 2000–2006. (Source: Calculated from estimates of per capita GDP in US dollars at current prices, data provided by the United Nations (30.1.08).)

Region	2000	2001	2002	2003	2004	2005	2006
Oceania	15,121	14,268	15,731	19,968	23,815	26,249	27,053
Europe	12,638	12,853	14,055	17,098	19,829	20,908	22,455
Americas	15,008	15,111	15,105	15,729	16,850	18,139	19,510
Asia	2,425	2,241	2,267	2,473	2,743	2,956	3,162
Africa	738	693	676	790	919	1,054	1,158
World	**5,201**	**5,101**	**5,243**	**5,822**	**6,465**	**6,896**	**7,372**
Subregion							
Northern America	33,245	33,849	34,622	36,180	38,476	40,742	43,124
Northern Europe	24,056	23,842	26,124	30,716	35,911	37,525	40,157
Western Europe	23,266	23,306	25,069	30,449	34,378	35,161	36,635
Australia and New Zealand	19,671	18,607	20,598	26,239	31,403	34,721	35,857
Southern Europe	13,751	14,099	15,530	19,378	22,342	23,238	24,705
Polynesia	6,907	6,801	7,452	9,247	10,578	11,056	11,456
Western Asia	3,960	3,550	3,726	4,238	4,997	5,948	6,826
Eastern Europe	2,128	2,434	2,748	3,347	4,300	5,351	6,505
Central America	4,790	5,050	5,210	5,110	5,410	5,996	6,412
Caribbean	4,221	4,349	4,465	4,523	4,783	5,362	5,788
Eastern Asia	4,648	4,280	4,266	4,627	5,094	5,369	5,584
South America	3,638	3,281	2,546	2,720	3,247	4,061	5,058
Southern Africa	2,761	2,442	2,268	3,365	4,308	4,760	4,829
Micronesia	1,897	1,883	1,904	1,995	2,081	2,191	2,251
Northern Africa	1,472	1,402	1,344	1,414	1,616	1,873	2,115
South-Eastern Asia	1,146	1,085	1,196	1,320	1,454	1,591	1,879
Melanesia	1,328	1,222	1,241	1,486	1,686	1,792	1,850
Middle Africa	362	355	395	467	585	764	936
South Central Asia	504	505	538	605	695	800	904
Western Africa	438	418	432	506	558	657	733
Eastern Africa	253	252	250	266	280	309	346

Construction work has been concentrated in centres such as New Delhi, Mumbai and other industrial centres, while rural areas, where the majority of the population lives, have had relatively little construction investment.

As the size of construction markets varies with income, the more rapidly an economy grows the faster construction markets might be expected to expand. However, while direct government intervention in China has stimulated the construction market, in India the construction market appears to have failed to have expanded in line with economic growth. This may indicate that there is a large pent up demand for construction work in India, which will increase the size of the Indian construction market in the years to come.

In more general economic terms, there are two scenarios arising out of economic growth which impact differently on construction. If an economy

Table 8.3 Regional GDP per capita as a ratio of the world average 1970–2006. (Source: Calculated from estimates of per capita GDP in US dollars, data from the United Nations Statistics Division website.)

Region	1970	1980	1990	2000	2006
Oceania	3.05	3.30	3.38	2.91	3.67
Europe	2.17	2.47	2.78	2.43	3.05
Americas	2.79	2.30	2.47	2.89	2.65
Asia	0.26	0.35	0.40	0.47	0.43
Africa	0.29	0.40	0.19	0.14	0.16
World	1	1	1	1	1
Sub-region					
Northern America	5.40	4.44	5.34	6.39	5.85
Northern Europe	2.87	3.98	4.44	4.63	5.45
Western Europe	3.10	4.59	5.21	4.47	4.97
Australia and New Zealand	3.69	4.06	4.28	3.78	4.86
Southern Europe	1.67	2.31	3.26	2.64	3.35
Polynesia	0.86	1.31	1.62	1.33	1.55
Western Asia	0.62	1.36	0.72	0.76	0.93
Eastern Europe	1.73	1.17	0.69	0.41	0.88
Central America	0.75	0.95	0.63	0.92	0.87
Caribbean	0.78	0.75	0.66	0.81	0.79
Eastern Asia	0.36	0.48	0.70	0.89	0.76
South America	0.64	0.75	0.62	0.70	0.69
Southern Africa	0.83	0.96	0.68	0.53	0.66
Micronesia	0.55	0.36	0.37	0.36	0.31
Northern Africa	0.32	0.47	0.31	0.28	0.29
South-Eastern Asia	0.14	0.20	0.19	0.22	0.25
Melanesia	0.47	0.47	0.33	0.26	0.25
Middle Africa	0.20	0.20	0.14	0.07	0.13
South Central Asia	0.15	0.14	0.11	0.10	0.12
Western Africa	0.32	0.60	0.14	0.08	0.10
Eastern Africa	0.16	0.12	0.07	0.05	0.05

grows as a result of increased productivity, then the growth can be used to improve the standard of living and to invest in capital or building projects, leading to increases in construction demand. Alternatively, if economic growth is the result of population increase then the GDP per head may even decline, if population grows faster than the economy. As a result, consumption spending may increase at the expense of investment spending, reducing demand for construction. This has been the situation in a number of countries, most notably in Africa, for example, in Zimbabwe. Although Table 8.4 shows that taking several countries together, the growth rates in all regions in Africa are greater than the world average.

Table 8.4 Average rates of growth 2003–2006. (Source: Calculated from estimates of rates of growth of GDP (in per cent), from the United Nations Statistics Division website.)

Regions and subregions	Percentage growth rate
South Central Asia	7.87
Middle Africa	6.81
Western Asia	6.60
Eastern Europe	6.33
Northern Africa	5.88
South-Eastern Asia	5.86
Western Africa	5.70
Africa	5.47
Asia	5.33
Eastern Asia	4.65
South America	4.32
Eastern Africa	4.31
Southern Africa	4.14
Caribbean	4.13
World	**3.56**
Polynesia	3.54
Americas	3.31
Australia and New Zealand	3.30
Oceania	3.25
Northern America	3.21
Central America	3.15
Northern Europe	2.86
Melanesia	2.35
Europe	2.13
Southern Europe	1.78
Western Europe	1.21

Global construction markets

Crosthwaite and Connaughton (2008) estimated total global construction spending was approximately $4.6 trillion in 2007. This is largely in line with other estimates of the global construction market. According to Global Insight (2008), the global construction market was expected to be in the region of $4.8 trillion in 2008 having grown from $3.5 trillion in 2003, an annualised growth rate of 6.7 per cent.

Table 8.5 summarizes the size estimates of the most significant construction markets in 2007 according to Crosthwaite and Connaughton. The largest single market by far was the US. At just over $1 billion, construction in the US was approximately 23 per cent of the global market. The Japanese construction market was the second largest at over $700 billion or approximately 17% of the global construction and China at $550 billion was 12 per cent. The European countries of Germany, Italy, France, the UK and Spain together were

Table 8.5 Size of construction markets in specific countries in 2007. (Source: Based on Crosthwaite and Connaughton, Davis Langdon Management Consulting, 2008.)

Spending in $ billion	Country
More than 800	USA
600–799	Japan, China
400–599	–
200–399	Germany
100–199	Italy, France, UK, Brazil, Spain, Korea
Less than 100	Mexico, Australia, India

approximately $1,000 billion and comprised 21 per cent of the global market. The remaining 23 per cent (equivalent to the US market alone) was shared by the rest of the world, including the Gulf States, Russia, India, South America and Africa. The construction market in India was almost the same size as the Australian construction industry at just under $100 billion, in spite of its population growth being higher than China's and its economic growth matching. The main difference between the two countries can be seen in their political systems and the amount of direct government intervention the authorities are able to exercise.

While one might expect a tendency for smaller markets to grow faster than larger markets, China, one of the largest construction markets in the world, had one of the fastest growing construction markets in 2006, though the fastest was India, although of course in absolute terms it remains a far less significant market. Italy, France and the UK all grew at slower rates than Brazil, Korea and Mexico. This reflects the impact of major investments in construction on smaller economies as well as a global trend, which is gradually shifting the major construction markets from Europe and North America to the newly expanding economies, often referred to as the BRIC countries (Brazil, Russia, India and China). Nevertheless, by implication when the GDP per head of population in the BRIC countries reaches levels found in the developed world, their construction market growth rates are likely to slow down.

Construction demand also depends on expectations of growth. Crosthwaite and Connaughton (ibid.) estimated expected construction growth rates for a number of countries. Table 8.6 shows their findings, which imply that both China and India (but not Brazil) were expected to continue on a far more rapid construction expansion path than countries in Europe and North America. If these high rates were to continue indefinitely, eventually these countries would, as China has done, overtake the slower growing countries named in Table 8.6. In absolute terms the slow rates of growth are sufficient for the maintenance of high standards of living in the slow population-growing states of the developed world. Their construction markets remain relatively stable compared to the construction markets in rapidly expanding economies.

Table 8.6 Average annual growth of construction 2007–2010. (Source: Based on Crosthwaite and Connaughton, Davis Langdon Management Consulting, 2008.)

Average annual growth rate	Country
More than 8%	China
6–7.99%	India
4–5.99%	Korea
2–3.99%	Germany, France, UK, Spain, Brazil
Less than 2%	USA, Japan, Italy, Australia

Concluding remarks

Rapid growth cannot be sustained in the long term as material and skill shortages eventually arise, leading to inflation in building costs forcing a deceleration in the rate of expansion and crises in construction markets.

These construction crises also arise for the following reasons. The long run sustainability of rapid growth rates is eventually threatened by the law of diminishing returns. It becomes increasingly difficult to justify investment as returns on additional investments decline. As this point is approached, the countries with the highest growth rates will tend to slow down. As their growth rates decline, the effect is to reduce demand for buildings and hence demand for their construction industries.

Net investment is investment in additional capacity. As fewer extra buildings are required the net investment component in the construction market is reduced. The capacity of the construction industry then exceeds demand for its services and construction prices decline as a result. Building costs also decline, partly due to increased unemployment of building workers and spare capacity in construction firms and suppliers.

However, as the economy experiences increased levels of income as a result of growth, land prices are likely to remain high due to speculative demand. This means that although the price of property may be high, the share of development costs shift towards higher land prices and cheaper construction. The effect on the construction industry is a major restructuring as weaker contractors run into financial problems and are forced to merge or withdraw from the market. This in turn implies an increase in the concentration ratio in these construction industries with a few large firms dominating the different markets for main contractors and construction specialisms.

Population growth does not always lead to economic growth and economic growth does not always lead to the expansion of construction markets. Much depends on government policies. However, where governments are willing to intervene, countries have, in several cases, been able to make the transition from low income economies to medium income countries and the construction industry is a key element in achieving such economic objectives.

While it may be possible to see the economy driving construction demand, in some countries investment in construction might have the potential to create a multiplier effect where investment in construction projects is a major source of local employment. While this may be a short-term benefit of investment in capital projects, the long-term benefits of construction output arise out of the improvements in sustainable developments, transport links, the building of industrial and commercial facilities and the improvement in health arising from better housing, water supply and waste treatment. These benefits lead to increased productivity, efficiency and output, which are needed if the high standards of living in the West are ever to be experienced by the majority of people living in the poorest countries of the world. Construction is part of the solution.

Note

1 In Chapter 4, the 'construction sector approach' and the scope of the construction sector are examined in more detail.

References

Carassus, J. (2004) *The Construction Sector System Approach: An International Framework*. Report by CIB W55-W65 Construction Industry Comparative Analysis Project Group CIB Report. CIB: Rotterdam.

Crosthwaite, D. and Connaughton, J. (2008) *World Construction 2006–07*. Davis Langdon Management Consulting: London.

Global Insight Construction Services (2008) www.globalinsight.com/construction. Accessed January 2008.

Stern, N. (2007) *Stern Review: The Economics of Climate Change*. Cambridge University Press: Cambridge.

United Nations Statistics Division, (2008) http://unstats.un.org/. Accessed January 2008.

Chapter 9

Global construction markets and contractors

Christian Brockmann

Introduction

Global players need global markets. Companies that understand themselves as global players use a geocentric perspective for their business. They standardize their products or performances, they employ worldwide branding, and strive for economies of scale. Examples of global players are the Coca Cola Company, Microsoft, Toyota and many others. There is seemingly no way to associate the construction industry with such market behavior. Typically, construction markets are characterized as local, not global markets but still there are companies that do business in many countries around the world. To bring order to this confusion we can distinguish between five different markets (regional, national, international, multinational and global). In all of these markets specific goods (or services) are demanded from construction companies by clients.

In ensuing sections of this chapter:

- *Construction goods* researches the nature of construction goods and designates them as contract goods, i.e. the terminology of the New Institutional Economics. It is important to first look at construction goods because their nature is very different from those of the manufacturing industry (exchange goods). Keeping this difference in mind, we will develop a very distinct path to globalization in the construction industry.
- *Geographical markets and competition* is dedicated to the description of the five different geographical markets and the type of competition in each of them. It uses archival data extensively. Initially, there remains the question whether we can assume the existence of a global construction market.
- *International projects* portrays international projects as large-scale engineering projects or as megaprojects. Smaller or midsize projects are dealt with by regional or national contractors and are generally not procured internationally.
- *Contract prices* reviews the pricing path from submission (market price) to contract signature (ex-ante contract price) and to project closure (ex-post contract price). It will become clear that this can differ considerably.

Managing this estimating process greatly influences the project success from the contractor's side.
* *Global players in construction* summarizes the previous information and insights to determine the characteristics of a global player. These are companies that can set up a network of procurement, production and delivery in any part of the world. To do so they revert to standardized processes while delivering customized products.

Markets are defined as economic (not physical) places of transactions where prices are formed through the interaction of supply and demand for a good (Cooke, 1996). This theoretical observation aside, there is no simple and practical definition of a market. Industrial organization theory approaches the problem from the perspective of goods by defining them as a bundle of characteristics, such as quality, location, time, availability, consumers' information etc., and then differentiates markets according to a limited subset of characteristics of these goods (Tirole, 2000). Regulation practice in the US follows the 5 per cent rule to determine a market. Here, it is assumed for a theoretical market that all suppliers raise their prices for one good by 5 per cent. If, subsequently, the profits increase, there exists a market for this good. Otherwise, there are some substitutes competing in the same market (Taylor, 1995). A third view is proposed by Porter (1985), who determines markets by the competition which he sees as a function of suppliers, substitutes, competitors and consumers. A fourth, very pragmatic, alternative is to look where consumers buy certain goods or where certain suppliers offer their goods. We will take this approach to market definition from the perspective of suppliers (i.e. contractors).

So, the question for the construction industry becomes: What different marketing strategies do construction companies employ that allow us to define separate markets? This perspective will enable us to differentiate between five markets with different geographical extents and to determine the skills needed for each market. Since the highest market level is the global market, this allows us to define the characteristics of global players in construction.

Macroeconomics and microeconomics as well as marketing provide the tools for the following analyses.

Construction goods

The idea of this chapter is to provide a foundation for the claim that global players in construction must resort to strategies different from those in manufacturing. At the same time it makes tools available for a deeper analysis of markets. The terminology of the New Institutional Economics, with its distinction between exchange goods and contract goods serves to reach this goal.

The most advanced research on goods can be found in marketing, where a distinction is made between marketing for consumer goods, industrial goods and services (Kotler and Armstrong, 1989). Construction provides goods in all of these categories. Homes are consumer goods, factories are industrial goods, and many services are required for the manufacture of both, be it through engineering firms or construction companies. As such, this typical distinction of three types of goods is not illuminating for the construction sector.

A more seminal way to characterize construction goods is to use the terminology of the New Institutional Economics. Here, a main distinction is made between exchange and contract goods. Exchange goods are already produced at the time of purchase, contract goods are not. Another often used term in New Institutional Economics is that of a transaction. A transaction comprises both a contract and an exchange of property rights. Using this idea, we can characterize exchange goods as those where the signature of the contract and the exchange of property rights happen at the same time. In the case of contract goods, there is a distinct time interval between these two acts (Alchian and Woodward, 1988). It is evident that construction goods typically belong to the group of contract goods. Some contract goods can be standardized (i.e. car wash, transportation) and, in such a case, exhibit more the characteristics of exchange goods.

All transactions require a measurement of the qualities of the respective good (Barzel, 1982). Generally speaking, goods comprise three types of quality: search, experience and credence qualities (Darby and Karni, 1973). Search qualities can be determined when buying a product and, therefore, only exchange goods possess such qualities. Exchange goods in construction are materials and equipment. Ready-mixed concrete, for example, will be tested when delivered to the site before pouring. A certain set of characteristics is determined at this time. Experience qualities can be determined by a trial and error method when buying a good frequently. In the case of ready-mixed concrete, the compressive strength after 28 days is such an experience quality, based on the concrete mix and the experience with the producer. Finally, credence qualities are those that cannot be determined at the time of the purchase without doubt. A repair service on construction equipment can serve as an example to illustrate credence qualities in exchange goods of the construction industry.

A contract promises a future performance. As such, contract goods cannot possess search qualities at the time of signing the contract; this is an ex-ante perspective. In an ex-post perspective, however, construction goods have many search qualities. For a typology the ex-ante perspective offers more insight. Typically, construction goods do not have experience qualities since these are based on a high frequency of contracting between the same client and contractor. While this idea is researched as relationship marketing in construction, it is not often employed (Davis, 1999). Relationship contracting uses also another mechanism that is called the 'shadow of the future' which describes a causal link from the future (cause: termination of relationship) to the present (effect: no fraud). Therefore, contract goods are characterized by credence qualities. These are further distinguished with regard to the search qualities of their outcomes. Contract goods that ex-ante have no search qualities but do so ex-post, possess quasi-credence qualities (i.e. construction goods). Those, where even the outcome has no search qualities have true credence qualities (i.c. medical services). Table 9.1 summarizes the above ideas.

The categorization in Table 9.1 is in a top-down order. Ready-mixed concrete has ex-ante and ex-post search qualities. It does not matter whether it also has experience or credence qualities. Relationship contracting cannot rely on ex-ante search qualities, but on experience qualities stemming from the ongoing frequent business relationship. Construction goods are typically quasi-credence goods and as such they are ex-ante intangible and ex-post tangible. The outcome of a project management contract is ex-ante and ex-post intangible, this makes an evaluation of the outcome difficult if not impossible. The tangible outcomes of projects are influenced by many parties and cannot clearly be attributed to the project management contract.

In this chapter, we are concerned about typical construction goods such as buildings or infrastructure. As quasi-credence goods, these are problematic since the extended transaction period offers a range of action to both parties involved before final delivery of the product. The root of the problem is asymmetric information and this determines a principal–agent relationship (Arrow, 1985). In such relationships, the principal employs an agent for a specific task and the agent possesses information that is not available to the principal. Opportunistic behavior is a possibility under these circumstances and must be expected. '. . . Opportunism refers to the incomplete or distorted disclosure of information, especially to calculated efforts to mislead, distort, disguise, obfuscate, or otherwise confuse' (Williamson, 1985). With opportunism as a possibility, there is a range of sources for uncertainty to the principal (see Table 9.2).

Hidden characteristics can be a problem in goods with search qualities,

Table 9.1 Typology of goods

Type of quality	Type of good			
	Exchange goods	Contract goods		
		Experience goods	Quasi-credence goods	True credence goods
Ex-ante search qualities	Yes	No	No	No
Ex-post search qualities	Yes	Yes	Yes	No
Experience qualities	Maybe	Yes	No	No
Credence qualities	Maybe	Maybe	Yes	Yes
Construction industry example	Ready-mixed concrete	Relationship contracting	Buildings, infrastructure	Project management

Table 9.2 Sources of uncertainty in market transactions

Sources of uncertainty		Ex-post (after signing of contract)	
		Principal can watch the agent's actions	Principal cannot watch the agent's actions
Ex-ante (before signing the contract)	Actions of agent are fixed	• Hidden characteristics • Adverse selection	
	Actions of agents are open	• Hidden intention • Hold up	• Hidden action • Moral hazard

be it ex-ante or ex-post (Stigler, 1960). The warranty period in construction safeguards the client (principal) to a certain degree. Yet, the uncertainty is not limited to products, it also extends to processes. Here we need to distinguish between the ability to perform and the willingness to perform. Only the ability to perform (through the prequalification of a search quality) can contain ex-ante hidden characteristics that are not easy to detect by a client. Adverse selection is a result of hidden characteristics when a principal misjudges the true qualities of an offer.

Hidden intentions and hold-up describe both the same phenomenon. The agent uses an opportunity in the contract to gain a one-sided advantage that counters the intent of the contract (Alchian and Woodward, 1988). While hidden intentions have been formed in the past, the hold-up is pursued in the present. The principal can draw conclusions from the agent's actions and detect some hidden characteristics and adverse selection or hidden intentions ex-post, while able to act against hold-up immediately.

Hidden action and moral hazard describe situations where the agent lacks incentive to take care of the principal, i.e. situations where he benefits from opportunism (Varian, 1999). They are most dangerous to the principal because the agent can act as he wishes without the principal becoming aware of being cheated.

The characterization of construction goods as contract goods – or more exactly as quasi-credence goods – draws our attention to the fact that global players in construction must pursue different strategies than those producing consumer goods. Ex-ante there are no products available for branding or marketing, there is no possibility to standardize goods and subsequently there are no economies of scale. So the question remains at this point: What distinguishes a global player in construction?

The sources of uncertainty prevail in all geographical markets but have different importance, and clients in different markets might choose different approaches to deal with them. The construction companies must have an answer for these approaches. As such the sources of uncertainty help to differentiate the markets.

Geographical markets

The construction industry has in principal the choice between five different market strategies: regional, national, international, multinational and global. For each market strategy, different characteristics can be identified for the company's orientation, its behavior, the segments for which it produces goods and for the type or organization. A synopsis is given in Table 9.3. In Table 9.3 the global market strategy is missing. The reason is that, at this stage, there is no clear view of how to fill the cells. This information will only be developed in later parts of the chapter.

Regional markets and competition

Construction goods have mostly (with the exception of prefabrication) the character of real property and this implies construction activities on a regional level. Many interactions between client and contractor require their presence on site. Materials are bulky and transport is expensive, so a network of suppliers around the site is a competitive advantage. Construction activities are concentrated around civic or economic centers. For these reasons, construction is

Table 9.3 Options for market strategies in construction (Meffert and Bruhn, 2000)

Market strategy	Orientation	Behavior	Segments	Organization
Regional	Local network	Local	Few segments, specialized	Local headquarters
National	Many local networks within the home country	National	Many segments, diversified	National headquarter, local subsidiaries
International	Many local and some international networks	Ethnocentric	Many national segments, few international segments, diversified at home, specialized abroad	National headquarter, local subsidiaries, overseas department
Multinational	Many local and many international networks	Polycentric	Many national and international segments, diversified	National headquarter, local subsidiaries, overseas department, international acquisitions

centered around local markets and all construction companies must have a regional market strategy. An example of a regional market is the city of Bremen in Germany. Typical segments of this market are general building, general civil works, harbor construction and foundation works. A more detailed analysis will find even more submarkets. The concentration of works around an economic center can be seen in Figure 9.1, where empirical data for a representative set of construction companies are displayed. For this market around 60–70 per cent of the activities take place within a radius of 25 km. In times of the deepening recession in Germany from 1995–2000 and further to 2005, markets expanded in geographical extension. Since the home market does not offer enough work, companies venture further away despite additional transportation costs and unfamiliar markets.

Market extension for the submarkets in harbor and foundation construction also occurs. The higher capital investment in these sectors cannot be financed on a small market – there simply is not enough work. Data from several large public clients (demand side) support the notion of a concentrated regional market. Very small and large companies are more concentrated than mid-size companies. While the small ones are lacking information about, and contacts in, other markets, the large companies delineate the activities of their subsidiaries through internal boundaries as a matter of organization (Brockmann and Nolte, 2007).

In Germany, in 2000 (in the middle of the recession) there were roughly 80,000 construction companies. Of these 78,500 had fewer than 100 employees. If we accept this figure to be indicative of local companies, then 98 per cent of all companies on this national market are local companies (Hauptverband der Deutschen Bauindustrie, 2003). The reasoning for this cut-off point is as

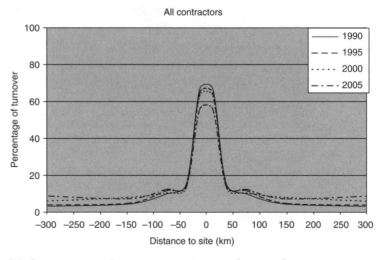

Figure 9.1 Concentration of construction activities in Bremen, Germany.

follows: a company must sustain somewhat substantive construction activities in a number of places to be called active on a national market. Ten regional markets with ten employees in each market seem to be a minimum expectation for a nationally active company. The market in the UK is especially fragmented, with a total of more than 164,000 companies in 1999. Out of these, approximately 750 had more than 100 employees, i.e. more than 99.5 per cent of the companies had fewer than 100 employees (Morton, 2002). In the US, there are more than half a million construction companies in the market and this market is also highly fragmented (Barrie and Paulson, 1992). So what applies to Germany is also true in the UK and the US. The fragmentation of regional markets can be explained by the demand. In 1999, 47.3 per cent of all new build work in the UK was orders with a magnitude of less than £100,000, yet only about 50 per cent of all contracts were new build work. Contracts for repairs and maintenance will be even smaller (Morton, 2002). Any industry sector with substantive numbers of contracts ranging from a few thousand euros to megaprojects worth several billion euros will show similar patterns of fragmentation.

Small companies are evidently more cost-efficient in dealing with small contracts. An explanation is given in Figure 9.2. Small companies rely on small investments, thus their fixed costs are small, accordingly unit rates are relatively high (i.e. €/m² floor area in housing). For large companies the opposite is true. Figure 9.2 shows the case of two companies, for clarity, but the facts do not change on the aggregate level of macroeconomics. Below a critical contract amount ($A_{critical}$), smaller companies produce at lower costs.

In general, competition in regional construction markets will be high and dynamic due to the fragmentation. Table 9.4 gives us data for the number of

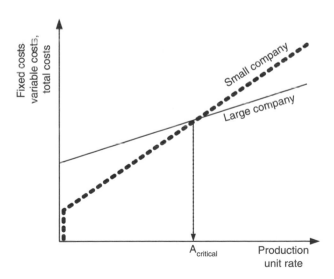

Figure 9.2 Production costs of small vs. large companies.

Table 9.4 Supply and demand for general contractors in Bremen, Germany (Bauindustrie-verband Bremen-Nordniedersachsen, 1997–2002)

Year	1993	1994	1995	1996	1998	1999	2000	2001
No. of buyers (housing)	739	746	482	565	817	922	925	751
No. of buyers (building)	142	116	126	119	211	209	203	151
No. of sellers	~130	~130	~130	~130	~130	~120	~120	~120

Table 9.5 Percentage of identical competitors in submissions (Nolte, 2006)

Percentage of recurring competitors	ca. 80	ca. 60	ca. 40	ca. 20	<20
Percentage of submissions	27	33	30	6	4

clients and general contractors in housing and building in Bremen, Germany for a number of years covering a business cycle. The interpretation is clear: in such a market there is all the evidence of perfect competition.

Collusion is always a possibility, if there is a closed group of bidders for specific types of tenders. Table 9.5 again gives data for Bremen. In 27 per cent of the submissions there are still 20 per cent new competitors. This provides the necessary dynamics to make collusion impossible. What these data do not prove is the absence of niche markets with very limited competition and the possibility of collusion. However those markets will be few in number and small in size.

National markets and competition

Once a company has established a regional presence it can add additional networks in other areas. At the end of such a process the company becomes a national competitor by being active in most of the important regional markets of a country. Since activities on the national market are the sum of regional activities, there can be no national strategy without regional activities. Since construction projects cannot guarantee continuous employment, most national companies diversify in order to ensure continuous production. National construction markets are defined by the building rules and regulations of the country as well as by the national language and culture. These form a barrier to entry.

National companies have headquarters and many local subsidiaries. Intra-company services are concentrated at the headquarters. The contact with the client is mostly looked after on the regional level although there are also some national clients. Other than the regional company, the national company has gained the competence to grow into other markets, albeit this is limited within the set of national culture and regulations.

While there are no data on how many national companies are active in such a market, it can be said with the same reasoning as above, that it must be less than 2 per cent of all companies. There will be variations from this level from country to country, yet the picture will not change dramatically. This does not mean that competition in the national market is limited. Regional and national companies fight in the same markets as was explained with regard to the market in Bremen. Large companies do not dominate the national market. The top five contractors in national markets do not share more then 10 per cent of the market in industrialized countries, with Japan being the exception at about 11 per cent (Bollinger, 1996). A way to measure concentration in a market is the Herfindahl-Hirschman Index (HHI). The HHI is the sum of the squared market shares in per cent. If there is just one supplier, this adds up to 10,000 (100 %²), if there are a thousand with a share of 0.1 per cent, then the HHI equals 10. Values of below 1,000 signify no concentration in a market (Stiglitz, 1997). Using data from the German Monopoly Commission in 2001, we can determine an approximate value for the HHI in Germany (Monopolkommission, 2003). Only an approximation is possible, since the data are given for classes of companies and not for single companies (see Table 9.6).

The HHI* has a value of 12.83 for construction in Germany, signifying a very competitive industry with no determinant players. The top ten contractors just have a market share of 9.1 per cent (Russig et al., 1996). Including more than 100 companies will increase the HHI slightly. The value calculated in Table 9.6 carries an asterisk to denote that it approximates the defined HHI.

The tobacco industry might serve as a comparative industry. Here the HHI* is 808.86 in Germany. Regional and national markets in construction are fragmented and market share is not a competitive advantage. This holds true for normal projects. Large scale engineering (LSE) projects are an exception on the national level. Regulators in Germany stopped the acquisition of shares from Ph. Holzmann (at that time No. 1) through Hochtief (at that time No. 2), reasoning that this would limit competition for LSE projects. This special case set aside, the number of construction companies in Germany and other

Table 9.6 Approximated Herfindahl-Hirschman Index for the German construction industry

Number of companies in a class	Market share of the class	Average market share of one company	Squared market share	Sum of squared market shares
3	4.8	1.60	2.56	7.68
3	2.0	0.66	0.44	1.33
4	2.3	0.58	0.33	1.32
15	4.7	0.31	0.10	1.47
25	3.9	0.16	0.02	0.61
50	4.6	0.09	0.01	0.42
			HHI*	**12.83**

countries (i.e. UK and US, see above) allows us to conclude that national construction markets in general show a very low degree of concentration and accordingly high competition.

The German construction market has experienced a long-lasting recession starting in 1995 and ending in 2005 (see Table 9.7). It will be shown, in a later part of this chapter, that economic activities on a national market have repercussions for all the following markets: international, multinational and global. Therefore, an analysis of the business cycle in national markets is of importance for global markets. Despite two moderate increases in construction spending in 1999 and 2004, there was a decrease from the high in 1994 to the low in 2005 by 23 per cent. How did the market react?

To explain the reaction of the market, a supply and demand model can be employed. In Figure 9.3, a shift in demand is shown. The quantity difference can be given as (264–251 =) 13 billion euros or 5 per cent. The price difference cannot be as easily determined since it is an aggregate figure from very different projects such as homes, bridges and tunnels. The reduction in the quantity demanded puts a very strong pressure on the prices to which the companies must react with time. Theory tells us that the available production factors such as labor, plant and equipment must be reduced. And indeed the number of insolvencies of construction companies increased from 1994 (ca. 2,800 per year) to 1996 (ca. 4,200). Then it stayed rather constant at a rate of 4,400 per year until the end of the recession in 2004.

Since there is an increase in insolvencies, one could expect to find fewer construction companies in the market, reducing the pressure on the price by a lesser competition. However the number of construction companies increased from 73,853 to 76,720 over a 10-year period with a sharp reduction in the number of workers (see Table 9.8).

While the number of companies has increased from 1995 to 2004, there was a shift from large to small companies. The larger the company size in 1995, the greater was the percentage of companies leaving the market in that group. The table shows also a markedly high decrease in labor as one production factor. Altogether, there was a shift from larger to smaller companies. The explanation is found in Figure 9.2: given small quantities demanded in a market as occurs during a recession, small companies produce at lower cost and can therefore better adapt to pressures on the price. The pressure on the price stays high since the number of competitors is not reduced – quite the contrary. The price index for residential buildings for general contractors (with a value of 100 per cent in

Table 9.7 Construction spending in Germany 1994–2006 in billion euros (Hauptverband der Deutschen Bauindustrie, 2003, 2005)

Year	1994	1995	1996	1997	1998	1999	2000	2001	2002	2003	2004	2005	2006
Billion €	264	259	251	248	245	249	242	228	214	208	211	202	217

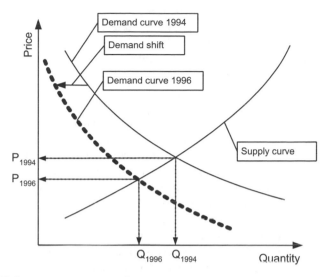

Figure 9.3 Shift in construction spending.

2000) fell from 104.9 per cent in 1995 to 97.7 per cent in 2003 (Hoffman, 2006). However, there were also other construction sectors where prices increased. Residential buildings are illuminating because these are the types of buildings that small companies build with preference.

To sum up the ideas above, it can be stated that during recessions smaller companies have a competitive advantage driving larger companies out of the market. The number of companies that can be considered as national construction companies decreases as well as the base for international, multinational or global construction companies. Competition in national construction markets is high and increases during recessions.

Opportunistic behavior is displayed by the client during the contract negotiation process taking the form of hidden action or hold-up. The contractor pays

Table 9.8 Company size and employees from 1995 to 2004 in Germany (Hauptverband der Deutschen Bauindustrie, 2005)

Company size (number of employees)	Number of companies 1995	Number of companies 2004	Change (%)	Employees 1995	Employees 2004
1–19	57,216	68,812	+20	391,557	361,517
20–49	10,866	5,578	–49	328,584	165,334
50–99	3,575	1,539	–57	246,305	104,707
100–199	1,524	584	–62	207,342	78,792
200 and above	672	207	–69	259,658	75,593
Total	73,853	76,720		1,433,446	785,943

back a phase later – during construction – using the same coinage. During negotiations he tries to convince the client to select his own offer, regardless of whether it might mean an adverse selection from the client's point of view. These tactics are deeply embedded in the culture of the country.

International markets

An international strategy is characterized by a large share of the revenues being generated in the home market and some additional activities in selected foreign markets. Construction services offered on foreign markets are specialized and limited to a small range. The number of countries served is also small. The behavior of the company is ethnocentric with a limited amount of knowledge of foreign markets. The international activities are coordinated from the national headquarters through an overseas department. Despite all limitations, international companies must have the know-how to form and manage the networks around their foreign sites responding to differences in culture and regulations. There is no scientific way of determining what is meant by a small number of countries. For reasons of clarity we propose here a cut-off point of less than ten countries.

There are few figures available for international activities but, in 1993 in Germany, there were as many construction companies as in 2000 (ca. 80,000) and of these 249 were active in international markets, which is equivalent to 0.3 per cent. Thirty-five companies (or 14 per cent of 249) had a share of 90 per cent of this market. The volume of international contracts was a mere 1.5 per cent of the volume of the German market. The conclusion is that very few of all regional construction companies are interested in and capable of leaving their home market. Many of them do the odd job just across a frontier and some take a deeper interest in international markets (Russig et al., 1996).

A look at the international revenue of the top six German inter- or multi-national construction companies allows us to estimate the size of each of these groups (see Table 9.9). Without looking at the organization of each company, it is hard to determine which of these companies are international and which are multinational companies.

These six companies are the only German companies in the ENR's list of the top 225 international contractors in 2005. The position 225 in this list is held by a company with an international revenue of just US$24.8 million, hardly a multinational. Therefore, it is safe to conclude that the list includes all German multinational contractors. A look at the structure of Heberger Bau allows us to determine whether it should be qualified as an international or multinational contractor. The company started international construction in 1974 and has subsidiaries in ten countries with a substantial international revenue (US$201 million). So it might qualify as a multinational using our criterion of activities in ten or more countries (www.heberger.de, 2007).

Looking at the six biggest international contractors in 1994 reveals how

Table 9.9 The top six German international contractors in 2005 (Reina and Tulacz, 2006)

International rank	Company	International revenue (million €)
1	Hochtief	17,599
9	Bilfinger Berger	6,553
63	Züblin	730
77	Bauer	442
89	SKE Group	366
116	Heberger Bau	201

Table 9.10 The top six German international contractors in 1994 (Bollinger, 1996)

Domestic rank	Company	International revenue (million €)
1	Philipp Holzmann	2,030
2	Hochtief	1,823
3	Bilfinger Berger	1,614
4	Strabag	565
5	Dyckerhoff & Widmann	304
6	Walter Bau	197

much influence a national recession has on contractors working internationally (see Table 9.10). A comparison between Tables 9.9 and 9.10 reveals that out of six companies in 1994 only two remain in the top ranks after 10 years of recession. Philipp Holzmann, Dyckerhoff & Widmann and Walter Bau declared bankruptcy. Strabag was bought by the Austrian Bau-Holding and is no longer a German contractor. It had, in 2005, an international revenue of US$10,799 million under the name of Strabag SE. At the same time, two companies increased the volume of their international activities substantially (Hochtief, Bilfinger Berger). All this evidence points to the fact that national markets are the base for international activities.

However, it is also possible to become almost independent of the home market. A good example is here is Hochtief who produces 88 per cent of its revenue on international markets. These data for all the top ten international contractors are given in Table 9.13.

Multinational markets

If the activities in foreign countries contribute a larger share to a company's revenues and if these contributions come from many different foreign markets (more than ten), then the company can be called multinational. International and multinational markets cannot be distinguished by groups of contractors. While contractors have either an international or a multinational strategy, they compete in the same physical markets.

The companies that are multinational differ quite significantly from those who are international. Ethnocentricity is replaced be a polycentric behavior with strong orientations towards the different host countries. This, of course, implies a decentralized organization of the company with many top managers in the host countries having a local origin. Besides coordination of the international activities through an overseas department, the company will acquire foreign companies. A multinational company has many foreign subsidiaries. See Table 9.11 for details of the German company Hochtief (Hochtief, 2003). The ability to work in foreign environments and to build up networks in a score of different cultures is the main competence of the multinational company. Through its polycentric behavior, such a company is prepared for and at ease in foreign environments.

Much of Hochtief's international revenue is, however, produced locally. Turner was the third largest contractor in the US in 2007, adding up most of its revenue in local projects all over the US. These projects were managed and staffed by Americans; the clients were Americans as well as the suppliers and the equipment producers. Technology and culture were also American. Since Turner is owned by Hochtief all these revenues count as international revenue for Hochtief. The rankings of the Engineering News Record for the biggest international contractors are based on the international revenue as shown in the books of the companies. Most of it is regionally produced with no foreign influence except the transfer of profits and an occasional meeting of top representatives.

Global markets

Global companies producing exchange goods (consumer goods) are not simply active in more foreign countries than the multinational company but they use

Table 9.11 Affiliated companies of Hochtief (Germany)

Continent	Affiliated company	Country
Africa	Concor Limited	South Africa
America	Hochtief Argentina S.A	Argentina
	Hochtief do Brasil S.A.	Brazil
	Aecon Group	Canada
	The Turner Corporation, Kitchell Corporation	USA
Asia	Leighton Asia Ltd.	Hong Kong
Australia	Leighton Holdings	Australia
Europe	Hochtief AG	Germany
	Hochtief (UK) Construction Ltd.	UK
	Hochtief Luxembourg S.A.	Luxemburg
	Hochtief Polska Sp.z o.o.	Poland
	Hochtief Russia	Russia
	VSB a.s.	Czech Republic

standardization of their products to achieve the highest possible economies of scale. They profit from a network spanning the globe to organize production. Managers come from countries around the globe, products are bought where cheapest, and production is set up where labor costs are low. The orientation of the company is not tied to any one national culture, it is global, the behavior is polycentric. The headquarters coordinates the activities of a multitude of affiliated and owned companies around the world. The important point for this strategy to work is, however, that the products can be standardized. It is out of the question that the products of the construction industry can be standardized around the world. So, global markets seem to be no option for this industry.

The five described markets form a hierarchy. Any multinational company is still rooted in a multitude of local markets, it is active in a number of national markets. One of these is the home market, the others are foreign or international markets. Projects and companies in many foreign as well as one home country characterize the multinational company. Companies need time to develop from a local contractor to a multinational contractor or a global player. For each step on the market ladder upwards, additional resources are required (see Figure 9.4).

It can be observed that companies climbing up the market ladder might lose some of the competencies they used to have. A multinational contractor will not only be not competitive in the small local market for housing, he will not even have sufficient know-how to build family homes. In other words, companies that can build state-of-the-art tunnels are not necessarily able to build small homes. This is symbolized in Figure 9.4 by blocks that do not reach down to the level of local markets. It has been stressed before how much multinational contractors depend on local markets and this can be seen in Figure 9.4 by an overlap between the resources required for local and multinational markets.

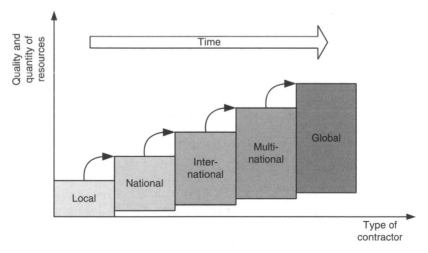

Figure 9.4 The market ladder in construction.

Table 9.12 Construction markets in Germany (2000)

Market	No. of companies	Percentage	Revenue (million €)	Percentage
Regional	75,220	100.00	78.6	100
National	>660	0.88	78.6	100
Multinational	~6	0.01	25.9	33

If one summarizes the data presented for the German construction market, then multinational construction companies are small in number. However, the revenue of the multinational contractors is by no means small. It amounts to 33 per cent of the total home market revenue (see Table 9.12). International and multinational construction is very concentrated from a national perspective.

Competition for international, multinational and global contractors

The total value of construction spending has been estimated to be around US$3.9 trillion for the year 2002 (Tulacz, 2000). In the same year the top 225 international contractors had a revenue of US$117 billion from overseas operations, which is equivalent to 3.4 per cent of construction spending. While nobody knows how much revenue is generated by smaller companies, it cannot be a big sum. The last five companies in that list only have an international revenue of US$1 million each. In 2002, construction revenue in Germany amounted to US$100 million, which means that the whole international market is roughly the same size as the market in Germany.

The interpretation of these data is clear: about 95 per cent of the global construction spending is allocated in local markets to local and national contractors. It definitely shows that the construction market is not a global market. Whatever happens in China is of very little interest to most construction companies in the US.

In Table 9.13, the ten companies with the largest international revenue in 2005 are shown (Reina and Tulacz, 2006). The sum of the international revenue of these ten companies amounts to US$99 billion which is 44 per cent of the worldwide international revenue (US$224 billion). Therefore we can conclude that the international market is very concentrated not only from a German national but also from a global perspective. Yet the market share of these ten companies of all national markets is merely 2.5 per cent. This percentage overestimates the market share, since total revenue is from 2002, while the revenue of the top ten is from 2005. This tells us that there are no global players in construction holding a recognizable market share. A closer look at the top ten companies shows different types of multinational companies. With the exception of Vinci and Bouygues all companies have larger international than national revenues. Technip has almost no home market.

Table 9.13 Revenue of the top ten multinational construction companies (2005)

No.	Company	Country	International revenue (bill. US$)	Total revenue (bill. US$)	International work (%)
1	Hochtief	Germany	17,599	19,795	88
2	Skanska	Sweden	12,347	15,722	78
3	Vinci	France	11,065	32,699	34
4	Strabag SE	Austria	10,799	13,502	80
5	Bouygues	France	9,576	24,960	38
6	Bechtel	USA	8,931	15,367	53
7	Technip	UK	8,084	8,245	99
8	Kellog, Brown & Root	USA	7,426	8,150	90
9	Bilfinger Berger	Germany	6,563	9,967	66
10	Fluor Corp.	USA	6,338	11,273	56

Table 9.14 The Herfindahl-Hirschman Index for international construction

Number of companies in a class	Market share of the class (%)	Average market share of one company (%)	Squared market share	Sum of squared market shares
3	18.3	6.1	37.2	111.7
3	13.1	4.4	19.0	57.0
4	12.7	3.2	10.1	40.2
15	20.9	1.4	2.0	29.2
25	16.5	0.7	0.4	10.9
50	18.5	0.4	0.1	6.9
	100.0	100.0	**HHI***	255.9

The HHI is calculated in Table 9.14 for the top 50 international contractors in 2005 in the same way as for Table 9.6. One small difference can be found. The total international revenue of US$224 billion has been allocated to the 100 contractors as shown in Table 9.14. This leads to a slight over-estimation in the HHI. The difference can be approximately quantified by looking at the group of 50 contractors. It is a fraction of the value of 6.9.

For comparison, the HHI* in 2002 was 324.8 so the international market has become more fragmented in the past 3 years. This is mostly due to the top three contractors, where, in 2002, the total value was 196 instead of 111.7 in 2005. Relative market share of the top three decreased.

To find out the strength of competition, we can look at different geographical markets. Anecdotal evidence from discussions with multinational construction companies is that they do not often compete for the same projects among each other. A reason could be that they prefer different markets or different types of projects. In Table 9.15, the top ten multinational contractors are grouped according to their revenue in six larger markets. The markets have the following size according to international revenue (US$ billion) in 2005: Europe (71.9), US (29.1), Asia (40.2), Africa (8.0), Middle East (41.4) and Latin America (15.9) (Reina and Tulacz, 2006).

Only in two construction markets, the US and Europe, are seven of the top ten multinational contractors active among the top ten for that market. Even in the next most important markets (Middle East and Asia), we find only three companies from the top ten. A differentiation of the top ten according to geographical markets can be found in Table 9.15.

The same picture emerges when comparing the different types of project that generate the international revenue (see Table 9.16). The idea is clear, there is not very strong competition among the top ten multinational companies that after all share 44 per cent of the market. The importance (US$ billion) of different types of projects is: building (59.5), industrial/petroleum (56.9), transportation (59.0), power (14.5), manufacturing (7.5) and water (0.85) (Reina and Tulacz, 2006).

Combining the geographical and type of construction information (cf. Tables 9.6 and 9.7), there are clear indications that, except for the markets in Europe and the US, the direct competition among the big multinational companies is

Table 9.15 Market competition in different regions (2005)

Middle East	Asia	Africa	Latin America	Europe	US
	Hochtief				Hochtief
				Skanska	Skanska
		Vinci		Vinci	
				Strabag	
		Bouygues		Bouygues	Bouygues
Bechtel		Bechtel	Bechtel	Bechtel	Bechtel
		Technip	Technip	Technip	Technip
Kellog		Kellog			Kellog
	Bilfinger	Bilfinger		Bilfinger	Bilfinger
Fluor	Fluor		Fluor		Fluor

Table 9.16 Market competition according to project type (2005)

Building	Ind./Petrol.	Transport	Power	Manufacturing	Water
Hochtief		Hochtief			Hochtief
Skanska		Skanska		Skanska	
Vinci		Vinci	Vinci		Vinci
Strabag		Strabag		Strabag	Strabag
Bouygues		Bouygues	Bouygues		
	Bechtel	Bechtel			
	Technip				
Bilfinger		Bilfinger	Bilfinger		
	Fluor				

rather limited geographically. The statement also applies to types of construction with the exception of buildings. A deeper explanation is also at hand: Hochtief, for example, participates through Turner in the building market of the US and, by being the international number two there, it has a considerable international revenue. Yet, the building market in all countries is large, fragmented, and local. No wonder the big ten are only in competition from time to time (see Table 9.17).

Table 9.17 is a possibility matrix, it does not show actual market activities, since it only combines information in Tables 9.15 and 9.16. If there were data on actual market activities, then the segmentation would be even stronger. Except for the building market in Europe and the transportation market in Europe and the US, a maximum of only three out of the top ten international contractors emerge in the matrix. The data support the statement that companies compete against a small subset of top international contractors. This again indicates that there is no global construction market, where companies compare themselves with the global leader.

Table 9.17 Market segmentation according to region and project type

Project type	Middle East	Asia	Africa	Latin America	Europe	USA
Building		Hochtief				Hochtief
					Skanska	Skanska
			Vinci		Vinci	
					Strabag	
			Bouygues		Bouygues	
		Bilfinger	Bilfinger		Bilfinger	Bilfinger
Ind./Petrol.	Bechtel		Bechtel	Bechtel	Bechtel	Bechtel
			Technip	Technip	Technip	Technip
	Fluor	Fluor		Fluor		Fluor
Transport		Hochtief				Hochtief
					Skanska	Skanska
			Vinci		Vinci	
					Strabag	
			Bouygues		Bouygues	Bouygues
				Bechtel	Bechtel	Bechtel
		Bilfinger	Bilfinger		Bilfinger	Bilfinger
Power			Vinci		Vinci	
			Bouygues		Bouygues	Bouygues
		Bilfinger	Bilfinger		Bilfinger	Bilfinger
Manufacturing					Skanska	Skanska
					Strabag	
Water		Hochtief				Hochtief
			Vinci		Vinci	
					Strabag	

International projects

The Engineering News Record (2000) has published a list of international projects and they claim that this list is rather representative for the market. In total, the list comprises 147 projects with an average size of US$401 million (see Figure 9.5). Even if the ten biggest projects are excluded from the calculation of the mean, the average project still has a volume of US$226 million. The conclusion is that a subgroup of international projects as demanded by the clients are, on average, large-scale engineering projects. Not on the list is the multitude of big or midsize projects in the category 'buildings' as described above for Hochtief.

Megaprojects still differ from international projects. Miller and Lessard (2000) have studied megaprojects around the world and the average contract size in their sample is US$985 million. As such, they form another special subgroup of all international projects. In addition, they require cutting-edge technology. Structures included in the sample are hydroelectric projects, thermal and nuclear power projects, urban transport, roads, tunnels, bridges, oil projects and technology projects. Not included, are buildings and manufacturing structures as in Table 9.16. Besides the technology, the companies involved in megaprojects need the ability to deal with the extreme complexity of such projects which includes the ability to establish the necessary network around the project in any part of the world. The network needs to incorporate a large number of stakeholders (see Figure 9.6).

International projects are enveloped by one group of international and another group of national stakeholders. Contractors, equipment producers, consultants and financiers are often international players, while subcontractors, material suppliers and authorities come from the domestic market where the project is located. The international contractor will often join forces with a

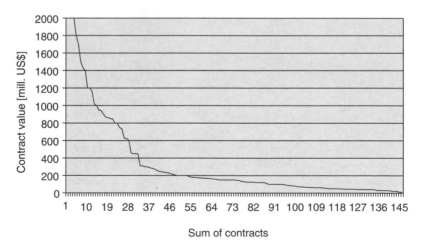

Figure 9.5 Contract values of typical international construction projects.

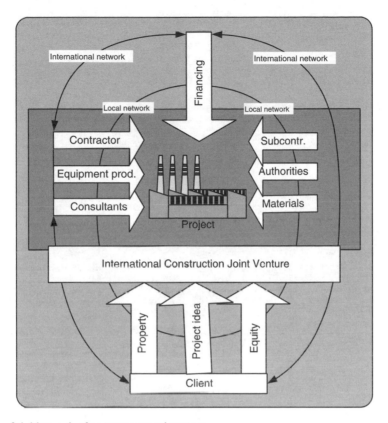

Figure 9.6 Network of an international project.

local contractor, since two groups must be organized. The international group will then be addressed by the international contractor, while the local contractor looks after the local group. Many international projects are of great public interest and the local contractor must then deal with such issues as well. To manage the network it is at least necessary for the international contractor to have joint venture experience. The network of main stakeholders for the BangNa Expressway in Bangkok is shown in Figure 9.7. This expressway is, at 55 km, the longest bridge in the world and a $1 billion megaproject.

As Figure 9.7 demonstrates, the BangNa Expressway is a multi-organizational enterprise (MOE). On the client's side, there are three ministries and two authorities directly involved: the Expressway and Rapid Transit Authority (ETA) as client and the Department of Highways (DoH) as owner of the right of way (RoW). The international construction joint venture (ICJV) comprised three partners, two German companies and one Thai company. Two of these (Bilfinger Berger and Ch. Karnchang) founded two companies as equity joint ventures (EJV) for the production of pre-stressed spun piles and post-tensioning

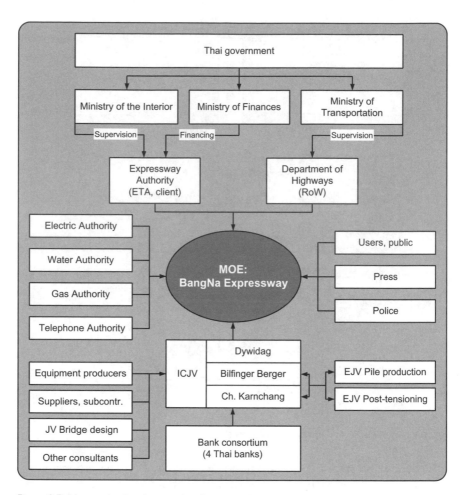

Figure 9.7 Network of an international megaproject.

systems. A multitude of equipment producers, suppliers and subcontractors provided the required resources as well as the Thai labor market, from which 5,000 workers were contracted. The design was produced on three continents and financed during construction through a bank consortium. Examples like these allow us to appreciate the know-how required to manage such a network.

Contract prices

As shown above, ex-ante and ex-post perspectives provide valuable insights for contract goods. This is also true for contract prices. Depending on the type of client, we can differentiate between processes with two or three phases. For the sake of simplicity, we assume that tender prices follow the rules of supply and demand. This assumption is somewhat problematic, because international construction markets are organized as sealed-bid auctions.

Auctions in construction are special, since the bidders are unable to estimate their costs exactly. This is most true for megaprojects that are expanding the technology frontier. We simply do not know precisely the costs ex-ante, since we have never gathered experience and collected cost data on a similar project. To know one's costs is called the private values assumption and this applies neither to LSE projects nor to megaprojects. If we accept, instead, that all competitors have the same costs (common values assumption) and make an estimating error ε_i, than the bid price will be $X_i = C + \varepsilon_i$. The individual estimate X_i is unbiased, i.e. on average equal to the expected costs. However, the low-bid is biased downward. The low-bidder must have an estimating error below zero, since the sum of the individual estimating errors is zero (Milgrom, 1989). If the mark-up is greater than the estimating error, then the contractor will not lose money. In the inverse case, he will not be able to cover his costs. This second case is not uncommon in competitive markets. Using these considerations, we would have to rethink microeconomic price theory that assumes sellers stop producing units when their marginal costs are equal to the market price. However, for the moment, we must accept economic theory as an approximation of the way in which market prices are determined in construction.

After the submission, the client knows the market price and he can start several rounds of negotiations to lower the price. In many countries, this is not an option for public clients. While all our data for the different markets indicate a market type that is close to perfect competition before submission, this does not hold any longer during the negotiation stage. Here, we find a monopsony where the buyer (client) yields considerable bargaining power because of highly asymmetrical information that only he can access. The case of monopsonies is not well researched because it is of no importance for exchange goods. We have no theory for the ensuing processes. The result of the negotiation phase is the ex-ante contract price.

Ex-post contract prices are established during the construction period and known at the end. The difference between ex-ante and ex-post contract prices is significant. In a study at Aalborg University, cost overruns for railway projects were at 45 per cent, for tunnels and bridges 34 per cent, and for roads 20 per cent (Flyvbjerg et al., 2003). In this phase, we find a two-sided monopoly with asymmetric information favoring the contractor. The result of the price path can be seen in Table 9.18.

The information in Table 9.18 applies to all geographical markets. However, in international construction the negotiation phase has a different quality and the same holds true for the construction phase. They are much more complex and culture plays an important role, especially during the negotiation phase. It is worth noting that the principal/agent relationship changes direction from the negotiation to the construction phase.

Table 9.18 Pricing path in construction

	Market type	Price path	Principal	Agent
Market price	Near perfect competition	Initial price	Does not apply	Does not apply
Ex-ante contract price	Monopsony	Lower than market price	Contractor	Client (more information)
Ex-post contract price	Two-sided monopoly	Higher than ex-ante	Client	Contractor (more information)

Global players in construction

Looking at the data presented in the previous sections, we can sum up the line of argument, as follows:

• Global players producing exchange goods market a standardized product around the globe, realizing economies of scale.
• We did not jump to the simple conclusion that construction companies can never become global players due to the nature of construction projects that can be characterized as contract goods and that are therefore customized (not standardized). Instead, we carefully described the market conditions.
• Contract goods have different characteristics from exchange goods. Opportunism is possible to a much larger degree, taking all the forms of hidden characteristics, adverse selection, hidden intentions, hold-up, hidden actions and moral hazard.
• Opportunistic actions are available to the agent in a principal/agent relationship. The role of the agent is not reserved for the contractor. During the negotiation phase, the client takes over the role of agent.
• Four markets open to construction companies can be easily described: regional, national, international and multinational markets. They are hierarchical and all activities are rooted in regional markets.
• While regional markets are highly fragmented, this is only true to a much lesser degree for international markets. On these competition is still dynamic, but not as much as in regional or national markets.
• Construction projects on international markets are either large-scale engineering projects or megaprojects.
• International construction projects require often the build-up of newly established and the management of highly complex networks.
• Market prices at submission and contract prices at the end of the project vary significantly.

When doing research on international megaprojects, we found convincing evidence that companies from very different cultures manage these projects in a very similar way. Existing literature tells us exactly the opposite. From this observation a case can be built for global contractors.

As a starting point we can look at the five preferred configuration of organizations as described by Hofstede and Hofstede (2000), who use the work of Mintzberg (1983). Hofstede and Hofstede have found five dimensions to describe different cultures (ethnocentric behavior). Two of them are used in Figure 9.8. A power distance index (PDI) describes the accepted degree of hierarchy in any one culture and an uncertainty avoidance index (UAI) measures how much ambiguity is tolerated and how it is dealt with. Mintzberg has identified five typical organizations: (1) simple structure with direct supervision; (2) machine bureaucracy with standardization of work processes; (3) professional bureaucracy with standardization of skills; (4) divisionalized form with standardization of outputs; and (5) adhocracy with mutual adjustment. These different organizational forms can be located on a grid of power distance and uncertainty avoidance. Using published literature, Hofstede and Hofstede were able to also allocate the different organizations as preferred

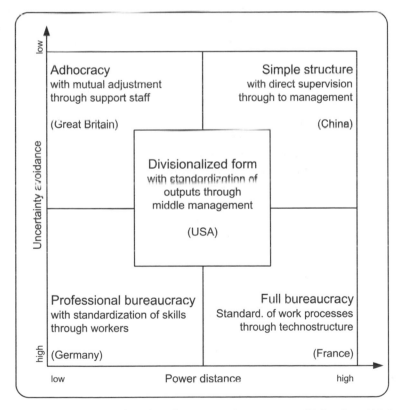

Figure 9.8 Five culturally preferred configurations of organizations (Hofstede and Hofstede, 2000).

forms to specific cultures. They found a match between the cultural values (PDI and UAI) and preferred organizational forms as discussed in the literature. This establishes a strong link between culture and organization.

Interview data from megaprojects very clearly give a different picture. Despite their national background and their cultural mix, all but one ICJV explained the necessity to form a decentralized organization. One ICJV differed by insisting on using their culturally preferred type – a simple structure with direct supervision – and the outcome was described by this ICJV and by others as a failure. All other participants had learned one lesson with time: for global megaprojects there is only one possible type of organization. It needs to be decentralized with empowerment of the middle management, with as much standardized skills as possible from the labor market (and little was found), with as many standardized work processes that needed to be developed in each case (since there are none in the beginning, there were just some at the end of the project), with a lot of adhocracy in the beginning and less at the end and with almost no centralized supervision.

After this introductory example of ICJV organization, there is the question of whether we can find further evidence of performance specifics for megaprojects. In a study of several infrastructure megaprojects in Thailand (expressway systems) and in Taiwan (high speed railways) we found indeed some striking evidence of globalization. There are five indicators of such specifics: the goals of the companies; the perception of complexity; the type of organization; the use of constructs; and technology. Interviews were conducted by the author with managers from the US, the UK, Germany, the Netherlands, Australia, Japan, Thailand, Taiwan and Korea (Brockmann, 2007). Regardless of the different national and company backgrounds, the interviewees described the overwhelming project complexity as the main characteristic of megaprojects. This corresponds with findings in the literature on megaprojects (see Miller and Lessard, 2000; Grün, 2004; Flyvbjerg et al., 2003).

Regardless of differences, they described their goals in unison as being mainly to generate a profit and to eliminate all obstacles that endanger this goal. Customer satisfaction is, for example, not seen as a goal *per se* but as a stepping stone to profit. Even Japanese managers point out that they will most likely work only once for this megaproject client and then customer satisfaction serves no purpose.

In spite of their cultural background, all ICJVs (with one exception) formed a decentralized organization and all worked as international construction joint ventures, where the international partner(s) manages mostly the international network and the local partner(s) mostly the local networks. Both relate with the client. The different interviewees held similar beliefs of their working environment. The constructs they used to put order to the world can be seen as cognitive maps and it is possible to draw such maps from interview data. They are shared by a large majority of the management.

Regardless of their construction experiences, ICJVs used similar cutting-edge technologies. Some had the know-how to do so without help, others needed help. The intermediaries were companies specializing in equipment (and running it) or in processes (and running them).

Summarizing the observations, we can state that there is far more than circumstantial evidence to support the idea that the construction of megaprojects is much more uniform than expected.

> The buyer generally doesn't know what he wants when he starts to buy it, no-one can actually be sure that what he requires can be produced, the production capacity to produce it doesn't exist at the time of commissioning and there are a large number of bodies and officials whose job it is to stop you getting what you want . . . In summary the building industry and its professionals sell a production service, not buildings.
>
> (Carpenter, 1982: p. 19)

This quote humorously describes the quality of construction goods. At the time of transaction they are complex contract goods that are described through incomplete contracts (Williamson, 1985). Construction companies offer their performance potential on markets and sellers look for indicators that signal just the required potential.

If it is true that construction companies sell their performance potential, another question with regard to global players in construction could concern whether there is a potential that is demanded by a global market. So the idea would be to no longer look at markets from the supply side but from the demand side. The answer to the question whether there is a global demand for a specific construction potential is yes and this with regard to international megaprojects.

All these indicators point in the same direction as the archival data. There is a select group of companies engaged in global projects around the world. They standardize across cultures and organizations what they offer on this market, their potential to execute megaprojects. Such companies are the global players in construction.

So finally we have to ask, what comprises this performance potential? By doing so, we can characterize global players in construction:

- A global contractor has understood the lessons of megaprojects, i.e. he knows how to deal with the overwhelming complexity. Such a company can be neither ethnocentric nor polycentric; it must transcentric in order to organize a megaproject: an organizational competence is required.
- Megaprojects are mostly built by ICJVs, so the global contractor must have experience with international joint ventures, especially since the local partner might never before have worked with an international partner and the same might hold true for the client: a cultural competence is required.
- The global contractor must organize and manage the international network: a social competence is required.

- The client must also (next to the local partner) deal with a foreign client, often with high-ranking government employees: a diplomatic competence is required.
- The international contractor must offer the client a worldwide reputation as hostage against his possibilities to resort to opportunism. Since the international contractor does not reside in the foreign country he can rather easily leave it. Once he behaves opportunistically, the client can only retaliate by hurting his reputation: a worldwide reputation is required.
- The global contractor must provide the cutting-edge technology: a technological competence is required.
- The global contractor must possess sufficient financial resources for the tendering, the negotiation phase and for the start-up of the megaproject: large financial resources are required.
- During the negotiation phase, the global contractor must understand the reversal of principal/agent relationship. He must find ways to withstand the pressures and to counter the asymmetric information favoring the client: a negotiation competence is required.

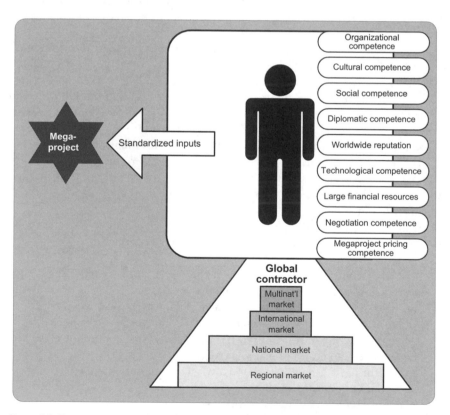

Figure 9.9 Experiences, competences, resources and behavior of global contractors.

- The international contractor must be able to price projects of a kind never before built: a megaproject pricing competence is required.

Considering the differences in the goods produced (exchange vs. contract goods), a global player does not standardize his output. To the contrary, he must standardize his input in order to deliver megaprojects around the world. The global contractor gathers the required competences and resources through his ascent from regional to multinational markets and beyond (see Figure 9.9).

The eLSEwise project undertaken in the 1990s comes with a very different reasoning to similar results describing success factors in the LSE industry: (1) ability to provide attractive financial packages; (2) ability to build winning alliances; (3) ability to accept and manage risk; (4) ability to invest in sales and R&D; (5) identification of client/user needs through market research; (6) ability to procure on a global basis; (7) technical expertise and the right technology; (8) integration of local and global knowledge; and (9) political backing (Male and Mitrovic, 1999). These results support the above findings.

References

Alchian, A.A. and Woodward, S. (1988) The firm is dead; long live the firm. A review of Oliver E. Williamson's the Economical Institutions of Capitalism. *Journal of Economic Literature*, **26**, 65–79.

Arrow, K.J. (1985) The economics of agency. In: Pratt, J. and Zeckenhauser, R. (eds) *Principals and Agents: The Structure of Business*. Harvard Business School Press: Boston.

Barrie, D.S. and Paulson, B.C. (1992) *Professional Construction Management*. McGraw-Hill: New York.

Barzel, Y. (1982) Measurement Cost and the Organization of Markets. *Journal of Law and Economics*, **25**, pp. 27–48.

Bauindustrieverband Bremen-Nordniedersachsen (eds.) (1997/2002) 32./37. *Öffentliche Vortragsveranstaltung*. Bremen.

Bollinger, R. (1996) *Auslandsbau*. In: Diederichs C.J. (ed.) *Handbuch der Strategischen und Taktischen Bauunternehmensführung*. Bauverlag: Wiesbaden.

Brockmann, C. (2007) *Erfolgsfaktoren von Internationalen Construction Joint Ventures in Südostasien*. Eigenverlag IBB: ETH Zürich.

Brockmann, C. and Nolte, L. (2007) Regional Construction Markets: the Example of Bremen, Germany. *Proceedings of the CIB World Building Congress, Cape Town 2007*, S. 3146–3158.

Carpenter J.B.G. (1982) Why build faster? – Commercial pressures. *Paper presented to Annual Conference, J.L.O., Oxford, Building Faster in Britain*, p. 19.

Cooke, A.J. (1996) *Economics and Construction*. Macmillan Press: Houndmills.

Darby, M. and Karni, E. (1973) Free Competition and the Optimal Amount of Fraud. *The Journal of Law and Economics*, **16**, 67–88.

Davis, P.R. (1999) Relationship Marketing in the Construction Industry. *1999 AACE International Transactions*, PM.11.1–PM.11.6.

Engineering News Record (eds) (2000) World market overview project list. *Engineering News Report*, Dec. 4, 45–52.

Flyvbjerg, B., Bruzelius, N. and Rothegatter, W. (2003) *Mega-projects and Risk: an Anatomy of Ambition*. Cambridge University Press: Cambridge.

Grün, O. (2004) *Taming Giant Projects: Managing of Multi-organization Enterprises*. Springer: Berlin.

Hauptverband der Deutschen Bauindustrie (eds.) (2003) *Die Bauwirtschaft im Zahlenbild*. Berlin.

Hauptverband der Deutschen Bauindustrie (eds.) (2005) *Die Bauwirtschaft im Zahlenbild*. Berlin. http://www.heberger.de. 27 November 2007.

Hochtief, A.G. (eds) (2003) *Geschäftsbericht 2003*. Essen.

Hoffmann, M. (ed)(2006) *Zahlentafeln für den Baubetrieb*. Teubner, Wiesbaden.

Hofstede G. and Hofstede, G. (2000) *Cultures and Organizations: Software of the Mind*. McGraw-Hill: New York.

Kotler, P. and Armstrong, G. (1989) *Principles of Marketing*. Prentice Hall: Englewood Cliffs.

Male, S. and Mitrovic, D. (1999) Trends in world markets and the LSE industry. *Engineering, Construction and Architectural Management*, 6, 7–20.

Meffert, H. and Bruhn, M. (2000) *Dienstleistungsmarketing: Grundlagen – Konzepte – Methoden*. Gabler: Wiesbaden.

Milgrom, P. (1989) Auctions and Bidding: A Primer. *Journal of Economic Perspectives*, 3, 3–22.

Miller, R. and Lessard, D. (eds) (2000) *The Strategic Management of Large Scale Engineering Projects*. MIT: Boston.

Mintzberg, H. (1983) *Structures in Fives: Designing Effective Organizations*. Prentice-Hall: Englewood Cliffs.

Monopolkommission (eds) (2003) *Hauptgutachten 2000/2001: Netzwettbewerb durch Regulierung*. Nomos: Baden-Baden.

Morton, R. (2002) *Construction UK: Introduction to the Industry*. Blackwell: Oxford.

Nolte, L. (2006) *Baumärkte als regionale Baumärkte – Nachweis am Beispiel Bremen*. Unpublished Master-Thesis, Hochschule Bremen.

Porter, M.E. (1985) *Competitive Advantage*. The Free Press: New York.

Reina, P. and Tulacz, G.J. (2006) The top 225 international firms are now more selective. *Engineering News Report*, Aug. 20, 30–45.

Russig, V., Deutsch, S. and Spillner, A. (1996) *Branchenbild Bauwirtschaft*, Duncker & Humblot: Berlin.

Stigler, W. (1960) The economics of information. *Journal of Political Economy*, 69, 213–225.

Stiglitz, J. E. (1997) *Economics*. Norton: New York.

Taylor, J.B. (1995) *Economics*. Houghton Mifflin Company: Boston.

Tirole, J. (2000) *The Theory of Industrial Organization*. MIT Press: Cambridge.

Tulacz G.J. (2000) A look at a $3.4 trillion market. *Engineering News Report*, Dec. 4, 30–43.

Varian, H. (1999) *Intermediate Microeconomics: A Modern Approach*. Norton: New York

Williamson, O.E. (1985) *The Economic Institutions of Capitalism*. The Free Press: New York.

Chapter 10

The new construction industry

Göran Runeson and Gerard de Valence

Introduction

This chapter deals with the creation of the 'new construction industry', a part of the construction industry that is as different from conventional construction as to constitute a separate, totally new industry. The driving force behind the development of this new industry is the tendency of large firms to grow larger aided by globalization and progress in communication technology. The result is a high technology oligopoly developing out of, but separate from, the traditional construction industry. It consists of a small number of very large firms operating in the global market, competing with technology and products, offering a complete project from material to design, finance, construction and operation. In this chapter we will first look at why and how firms grow, how the environment in which the building industry is operating is changing and how construction firms are responding. Finally, we will look at the theoretical and empirical implications of these responses in the form of the new construction industry.

Growth and the theory of the firm

Considering the importance of the growth of firms both for the continuing increases in the national income and the structure of industry, one would expect that this was a well developed area of theory, but nothing could be further from the truth. We know a little bit about what tends to happen. We can classify different apparent reasons and predict some outcomes. However, when we want to explain why it happens, there is a number of conflicting theories pointing, at best, to the inhibitors of growth, where they offer conflicting explanations.

There is a number of reasons why firms grow. They include increased profit, economies of scale, efficient utilization of plant and machinery, access to finance, expanding markets, long-run rather than short-run profit maximization or just 'men's (and women's) animal spirit'. There is a number of constraints on growth as well. These include the size or rate of growth of the market (Downie, 1958), and the access to investment capital without diluting ownership (Marris, 1964). Management can both inhibit and stimulate growth (Penrose, 1959). Management inhibits, or rather restricts, the rate of growth by the rate at which new management can become fully productive within the firm. Management may also stimulate growth, as when there has been growth in the past. The new management employed to cope with this growth will eventually become fully productive. Because of this, there will then be surplus management. The surplus management has experience, not only in management within the firm but in managing growth. They give the firm a potential for relatively painless growth at an accelerated rate.

Technological change, by creating new or better or cheaper products is an essential part of dynamic competition, which Schumpeter (1942) referred to as 'creative destruction'. It gives firms a competitive advantage when they introduce new technology and this enables them to grow. Sometimes the advantage allows a firm to perpetuate the advantage and grow into a dominant position, but sometimes the advantage is only temporary as competitors respond with their own innovations.

There are also different types of growth. The two most important are the creation of new capital and the acquisition of an existing firm through a takeover or merger (Runeson, 2000). Creating new capital normally requires the creation of a new or extended market with all the efforts that this involves. In contrast, acquiring an existing firm through a merger or takeover, means taking over an existing market as well. This reduces risks and saves resources and effort. A merger or takeover of an existing firm is, therefore, in many respects a much more attractive proposition than other forms of expansion.

Growth, by whatever means, may be horizontal, producing more of the same product, vertical when the firm engages in earlier or later stages in the production of its current output, or diversified when the firm moves into an unrelated type of production.

Eventually, the most successful firms grow out of their regional or national markets and move into new markets. Until 20 years ago, this was the beginning of the multinational firm. It started producing in different countries to escape various barriers to trade such as quotas, tariffs or transport costs. By locating branches in different countries, it avoided these barriers. Even better, it gained protection by the barriers, once it operated inside the national borders in the same way as a local firm would.

Now, however, when we have a global market where capital and goods and, to some extent, also people can move freely across national borders, we see a totally different type of firm, the global firm. It is not restricted by borders or barriers to trade. It can select freely the locations of the different functions of the firm on the basis of the availability and costs of the factors of production, in particular labour and skills, as national borders no longer matter.

The result is that most industries are dominated, or in the process of becoming dominated, by a small number of giant firms that satisfy a global market. In a conventional, comparative statics analysis, these firms represent allocative inefficiencies. However, in a dynamic setting, these firms are a strong force for progress as Schumpeter (1942) envisaged.

Construction and globalization

What is happening in other industries is also happening now in the construction industry, but because of the special characteristics of construction, the outcome is a little different. Essentially, there are no economies of scale in construction. Rather, there are diseconomies (Runeson, 2000). The sizes of firms are determined by the sizes of projects in the market. Effectively, the most competitive firm is the smallest firm that can muster the resources needed to carry out the job. Hence, there is little incentive for small or medium-sized firms to grow, unless they attempt to move into different markets. The growth of the very large firm has been determined by the increase in the number of very large projects or, lately, by projects that require the provision of new services in addition to construction management, most often some form of vertical integration.

For at least the last 80 years, there have been continuous calls to the construction industry to improve productivity in all sorts of inquests or reports (e.g. Cole, 1920). There is little evidence that any of these calls has had any effect. The changes in the building and construction industry have not depended, and will most likely not depend, on attempts from outside to influence the industry to become more innovative and more technologically advanced. Projects like the Latham (1994) and Egan (1998) inquiries and similarly motivated projects in other countries are interesting mostly because they are such curious anachronisms in their calls for government involvement, while we are living through an almost extreme shift to the right in all other areas of politics and economics. Rather, the future of the industry will depend on how the firms, and particularly the large firms in the industry, respond to changes in their environment. Some of these changes may be government initiated, others are technological, social or economic.

In this context, it is worthwhile looking at the term 'industry'. In the absence of a special definition, the term is at the same time too broad and too narrow to serve a discussion of the future of construction.[1] SIC-based definitions put multinational firms and self-employed craftsmen in the same industry, but exclude manufacturers of prefabricated building elements, architects or consulting engineers. To avoid confusion, for the rest of this chapter, we shall use the term 'construction industry' when talking about the conventional industry and 'construction' to refer to all activities involved in a construction project, from the production of building material to design, finance, construction and possibly also operation, maintenance and demolition. However, before dealing with the response from the industry, we will look at how the environment of the industry is changing, starting with the activities of the government.

Governments govern firstly by general social and economic policies to stabilize the level of economic activity, to ensure that markets work without impediments and to protect health, welfare and property rights. Secondly, they may promote or protect a specific industry. While the first is desirable, especially for an industry like construction, where economic disturbances are so destructive and which has such a large multiplier effect, the second is generally considered detrimental to good economic policy. The counter-argument is that, if an industry or firm can't survive without assistance, it is not making a reasonable contribution to the economy, and should be allowed to die in peace, so that the resources can be utilized where they give a higher return.

The idea, for instance, of a government-sponsored industry strategy to adopt innovations to be more productive or more like other industries, as often suggested, is based on an idea of the world so far from reality that the most basic economic aspects are ignored. Innovations and changes are introduced if, but only if, they add to profit. Much of the current discussions on innovation in construction have two things in common; they ignore nearly 100 years of economic research into the 'whys and hows' of innovation and there is no appreciation of the role of profit. Instead, they are based on the premise that if the technology is there, it should be adopted. Failure to adopt innovation is seen as resistance to change.

Innovations may be classified as product and production-process related. Profitability of product innovations requires control over design, so that the innovation can be put to use in the market, protection, so that it can't be used by firms that have not contributed to the development, and market power, so that the investments can be recovered. In the construction industry, where there is no control over design and no market power, the only innovations we will ever get are process innovations, mostly embodied in capital and therefore dependent on growth of the industry. There is rarely any competition to be the first to adopt process innovations based on common knowledge and pay the development costs. Firms adopt new technology after others have paid for the cost of developing them and mostly to defend existing profit rather than create new. Product innovations, if there are any, must be driven by the clients, who by and large are not interested. This has basically been the situation in the construction industry up until the last 10–15 years, but it is beginning to change. Australian firms are used here as examples, but similar developments are evident across the global market.

Despite the public perception and without government-sponsored strategies, there has been continuous progress in the industry in a number of areas. These changes are driven at least primarily by globalization that is affecting construction everywhere. Individually, the changes are hardly noticeable but, together, they have the potential to change fundamentally, both the industry and the way we look at construction, and we hope to justify this statement after briefly describing some of the new trends.

Like all major changes, economic, social or political, these changes have started a complex process of responses. It is not possible to cover every aspect of the ongoing development in one short chapter like this. Rather, the aim here is to describe the process in broad terms, indicating the reactions to the changes and their logical conclusions, with a case study illustrating the creation of the 'new construction industry'.

Parts of the preconditions that were brought into existence in Australia were the result of policies by the state and federal governments, as they, like the governments in many other countries, worked to remove any impediments to trade across the economy. This was necessary to derive the full benefits from the emerging global market. The construction industry in particular has benefited from this. Two state Royal Commissions and federal industrial relations reforms have helped to clean up the industry that was riddled by crime and corruption, facilitated by a chaotic industrial relations system overdue for change. The new industrial relations policy has created a new relationship between unions and employers and largely removed excesses on both sides. In this new system, individual contracts or enterprise bargaining take the place of awards imposed by quasi-legal tribunals.

In the upper end of the industry (the top 0.5 per cent or so of firms but accounting for more than 30 per cent of all work), this has led to a new appreciation of the employees as the ultimate resource of the contractors. Bonus schemes that give employees a share of the profit are lowering labour turnover at all levels, and helping to create conditions where employers are starting to invest in the training of their employees. This leads not only to a more skilled and flexible labour force but also to a softer management hierarchy, where decision making can be decentralized. Overall, there is a serious skills shortage in Australia, but a gap is opening up between the top end of the industry and the rest.

The state governments have also cleaned up the system of payments in the industry with Security of Payments Acts. The new Acts regulate payments within strict time limits that cannot be negotiated away in a contract, and the cash flow throughout the industry has improved. Before the Acts, payments were withheld regularly by contractors who could double their profit on a job by managing its cash flow. When such a contractor ran into trouble and went bankrupt, it could bring down with it 400 or 500 subcontractors, simply by the consequences of having withheld their payments. Now, any irregularities in payments will send out early warning signals that a contractor is no longer liquid.

Since the excesses associated with the financial deregulation more than 20 years ago, and the hard landing that followed interest rates at 22 per cent, there has been a reasonably steady economic growth with only a few minor hiccups. In the main, as we have seen, the government has changed the industry by removing obstacles to the efficient workings across all markets. They have also, for purely selfish reasons, done one more thing that has been instrumental in bringing about real change in the industry and that may in the long run be crucial to the development of the new construction industry. Both the federal and the state governments have participated in the world-wide trend towards the private supply of public projects (PPP). In such arrangements the finance, design, construction and operation are supplied by what is in effect a single supplier.

It is difficult to overestimate the differences between the roles of the contractor in a conventional project and a PPP. In a conventional project, the contractor provides a service. He/she competes on price to provide the management services required to construct a project designed by an agent for the client, financed by the client and handed over to the client on completion.

A PPP, on the other hand may be initiated by a client or a contracting organization – normally a syndicate – that can serve the functions of financier, designer, developer, contractor and facility manager. The organization competes on design, technology and value for money and designs, finances, constructs and operates the facility for anything up to 30 years. Rather than a fairly simplistic construction management service, the PPP contractor provides a sophisticated, differentiated product. Because of the product differentiation, the contractor gains the market power and, with that, the control necessary for recouping investments into the development of new technology and new products.

The major client for PPPs so far has been the public sector but private organizations are starting to follow their example. The benefits to the private sector are the same as for the public sector, but in addition, clients can utilize the experience of the contractor. Firms with experience and skills in things such as leisure facilities, retirement villages, health care or sustainable building have a competitive advantage. They can draw on these skills to create additional value for the client. They can deliver to a market that increasingly expects more services, higher quality and lower costs combined with environmental sensitivity.

While there may be diseconomies of scale in simple construction, for the kind of work we discuss here, where the organization serves all of these added functions, there are definite economies of scale and if we add research and development (R&D), it is likely that the economies of scale extend well beyond the size of any current firm. This provides a powerful incentive for the large firm to grow into a giant firm and to do this by diversifying around a core of construction activities. The easiest way to grow is through mergers and acquisitions and to become an international firm.

It is, therefore, not surprising that there has been a considerable consolidation in the top end of the industry, with mergers and acquisitions. Most, if not all of the large firms in any country now have an international connection and see themselves operating in a global market. There are more and more close links both up and down the production chain. The large and successful firms are diversifying both vertically and horizontally. Overall, the large firms are getting larger and the small firms smaller. Some medium-sized firms create niches for themselves while others are taken over or merge with large organizations. We'll see how this works in practice later in this chapter.

What has been described here is really a series of quite unremarkable changes, yet they are all part of, and together point to, a single, quite remarkable – almost revolutionary – change, and we will look at what that implies.

Fifteen years ago, David Hawk (1992) interviewed the managers of the world's largest firms involved in construction about their strategies and aims for the future. From these interviews, he identified a set of trends, which he referred to as the 'conditions of success'. These 'conditions of success' overlap almost completely the trends we have outlined here. Together, they point to a future with a small number of leading firms in the industry developing in the same direction. They may come from very different aspects of construction. They may have started as contractors, architects, material producers, subcontractors, consultants, developers or financiers, but they are now merging, acquiring each other and integrating vertically until they cover several, maybe all, stages from design and material production to operating or demolishing the building at the end of its service life (see also de Haan et al., 2002). At the same time, they are growing horizontally through the same means to acquire a market that can support all these functions. Very often, the mergers or takeovers are international to further extend the market. In this process, the number of firms involved has decreased drastically. This is demonstrated very clearly in Hawk's (2006) recent follow-up study. Of the 60 firms investigated 15 years ago, only 25 were still there as independent entities (see also, Seaden et al., 2003).

The conditions for success imply a new kind of management that reflects the need to manage change and flexibility. Formal and informal training of a stable work force become more and more important and so does technology. Most importantly, while all firms started from a strong domestic position, they have now grown out of the domestic market and operate internationally, aiming for a global market.

The new construction industry

As we have discussed here, this pointed to radical changes in how these firms functioned. The changes are so radical that, in his paper, Hawk referred to it as the development of a new industry, the new international construction industry.

Some of the trends Hawk saw have been slow to develop. The development in prefabrication has, for instance, been patchy. On the other hand, in Australia the employer–employee relationship has changed radically over the last 15 years, so much so that it is now an asset rather than something holding back development. Similarly, the use of PPPs and other innovative procurement methods has accelerated, we think, well beyond the expectations at the time, and mergers and acquisitions have been very frequent. We are moving, it seems, with a sense of inevitability, and without government involvement or any industry strategy, towards the development of this new industry. So, what will it be like and how will it be different from the 'old' industry?

This new industry, the logical conclusion of the changes we have listed here, would consist of a small number of giant firms, producing, not a service as the rest of the industry but a product in a business model that owes more to conventional manufacturing than to traditional building and construction. The consolidation will be ongoing. The firms will compete on value for money on procurement contracts similar to current PPPs. The CEOs would be on fixed-term contracts with remunerations determined by performance as measured by share prices and return on capital, and most of the employees will be on some form of incentive payments. The firms will attempt to retain their competitive advantage through R&D and develop global models of governance that maximize the advantages of the size and diversity of the firm.

The new industry is, to all intent and purposes, a modern, global manufacturing oligopoly, where a small number of large firms compete with their products in a global market. The products are packages, complete buildings, designed, financed, built, maintained, operated and possibly also demolished, as increasingly demanded by the clients, or rather customers.

The new construction industry is not going to compete with the main body of the traditional, very fragmented building industry that we know now. That part of the industry will continue to provide largely undifferentiated management services, allocated on price alone, to erect buildings designed and financed by the client, in a local or regional, or possibly, sometimes, even national market. It will continue as a low-technology industry in an environment where new technology is restricted to the process itself, and where product technology has to be financed and driven by the clients. The old and the new industries will be so far apart that it is possible there will not even be any technological spillover effects from the new industry.

The firms in the new industry are likely to continue to grow to exploit economies of scale and any new technologies they have developed. Partly, that growth will be in turnover, partly in geographical coverage and partly in new services and products as demanded by the market. Mergers and acquisitions are likely to continue to be a major strategy for generating this growth.

With a higher technology, the skills of the workforce will become more important. Firms in the new construction industry will be more interested in and take a much more active role in developing new skills. Having developed new skills, they will develop reward systems to make sure that labour turnover is reduced and skills retained within the firm. Experience with particular types of projects will become an important asset to the firm as the clients look to the producer to add extra value on specialized facilities such as those for sports, recreation, health or education.

One of the effects of changing attitudes between employers and employees is the softening of the hard management hierarchies that predominate in building, which has already started to increase efficiency. More decentralized decision making will increase flexibility and reduce the intensity of management and so reduce any remaining diseconomies of scale.

In the traditional industry, the contractor just constructs what someone else has designed. With more influence over the product, the new industry will have to demonstrate a more socially responsible attitude, for health and safety on site and especially for the environment. Construction and buildings now generate some 30–50 per cent of all greenhouse gases. If we envisage a world where everyone lives on the same standard of living as US, or for that matter Australia, our total emissions have to be reduced, not by 5 or 10 per cent as in the Kyoto agreement, or the 25–40 per cent agreed on in Bali, but by 90–95 per cent, and it is obvious that construction will have to play a major role. Ecological sustainability can no longer be ignored, nor can it be seen as an ethical issue only, that can be off-loaded on to others. It is more and more an area of new business opportunities and profit to firms that diversify into environmental services and demonstrate responsibility by developing environmentally sensitive solutions. This is strongly reflected in the way many, if not all, of the global construction firms now promote environmental responsibility through their mission statements.

From a small-scale Australian contractor to a global giant: some illustrations

Australia offers a good illustration to the thesis proposed in the first part of the chapter: that there is an ongoing process of mergers and acquisitions which is moving part of the industry towards being part of a small number of giant global diversified construction firms. Of the ten largest firms at the time when Hawke wrote his paper, only one has not been acquired by what is ultimately an international construction firm. The exception is Lend Lease Corporation[2], which instead has acquired, among other firms, the Bovis group, making it the world's largest construction management contractor with an established presence in the Americas, Europe as well as in Asia Pacific. With experience in areas such as master planning, concept design, value management, sustainable development, authority management and town planning, it is well on its way to become a diversified global giant.

Possibly more typical for the Australian industry, and equally conforming to the discussion above, is Leighton Contractors Pty Ltd, which is now part of Leighton Holdings Ltd. Leighton Contractors, which was founded in 1949, together with Theiss Pty Ltd (since 1983), formed the Leighton Group, one of the top ten firms in Australia. In the 1970s, it expanded into the Middle East, South, Southeast and East Asia through Leighton Asia Ltd and Leighton International Ltd. Both subsidiaries were established in 1975, but the current names date back only to 2007, after a restructuring in response to an increased interest in the Middle East. Leighton Properties Pty Ltd (1972) was established at the same time to cater for a diversification into property and development. The Leighton Group is also operating in Papua New Guinea, New Zealand and, recently, in South America.

Despite continuous growth, domestically and globally, there were some minor set-backs associated with falls in demand in the industry. So, for instance, an attempt to join the American market come to an end when, in 1993, Leighton Holdings sold the recently acquired Green Holdings, terminating its presence in the US. However, this was a temporary set-back only and Leighton Holdings continued to diversify, not only geographically, but in terms of activities. In 1996, it became a major player in Australia's telecommunications market with its purchase of Visionstream, and in 2006 the acquisition of the Australian–New Zealand contract mining assets of Henry Walker Eltin Group Limited further strengthened its position in the contract mining market.

In 2000, Leighton Holdings took a controlling interest in John Holland Group Pty Ltd (another top ten firm) and over the next 4 years it increased its stake to 99 per cent. While this was happening, John Holland added to its already broad contracting capability by acquiring Transfield Construction.

Recently, Leighton Holdings has further expanded its presence in the Middle East joining forces with Dubai-based Al Habtoor Engineering, a major construction firm which has more than 25,000 employees. Together they will form a new entity that will be called Al Habtoor Leighton which will certainly increase Leighton's activities in the Middle East but may also extend them into new areas such as North Africa.

The acquisitions, mergers and internal growth mean that Leighton Holdings is no longer solely a construction company. While construction remains its core activity it has active interests in engineering infrastructure, mining and resources, telecommunications, property and environmental services, and the type of jobs it aims for are projects where it can compete with value for money rather than price alone.

While this was happening in Australia, the ownership of the Leighton Group also changed. After some 20 years of having been the major shareholder of the Leighton Group, Leighton was acquired by the German construction firm Hochtief AG in 2001. At that time, Hochtief had recently expanded also into the US by purchasing Turner in 1999. Together, the two acquisitions provided Hochtief with a broad presence in North America, Australia and Asia as well as Europe, and it is now ranked number three in Europe. Hochtief, at the time it acquired Leighton Holdings, was controlled by RWE AG which was primarily involved in electricity supply and environmental services on an international scale, although it also had some interests in construction in addition to Hochtief. However, in 2004 it sold its shareholdings in Hochtief, more than 56 per cent, to institutional investors. After a period with no major shareholder, ACS (Actividades de Construcción y Servicios, S.A.), emerged in 2007 with a holding of just over 25 per cent, a controlling if not majority holding. ACS is a Spanish construction firm, similar in size to Hochtief, but so diversified that less than half of its revenue is derived from construction.

Hochtief was the largest in a series of mergers and investments that had created the ACS Group, which is another example of rapid growth through mergers and acquisitions. It began operating in 1983, when a group of engineers acquired and restructured Construcciones Padrós, a medium-sized construction company with financial problems. The strategy was repeated with the acquisition of OCISA in 1986, another construction company before ACS's first move towards diversification through the acquisition in 1988 of SEMI and the following year Cobra (electric and telecommunications companies).

The first in a series of large mergers took place in 1993 with the creation of OCP, which became one of the leading groups of the construction sector in Spain. The second large merger took place in 1997, with the creation of ACS as a result of OCP's mergers with Auxini and Gines Navarro. The group also diversified into its current services areas through acquisitions of Continental Auto (passenger transport), Onyx (environmental services), Imes (public lighting services, integral maintenance and control services) and Vertresa (the largest waste treatment plant in Madrid). The turn of the century started with the integration of the Dragados Group, making the ACS Group the most important construction group in Spain. Parallel to this, the ACS Group made strategic investments in Abertis (construction), Urbis, Unión Fenosa and Iberdrola (energy) as well as Hochtief.

Despite ACS's leading role in Spain, it has little experience in international contracting, but the acquisition of Hochtief will contribute to make it a major global construction firm. The combined revenue of the two firms as well as their combined construction revenue would make them number one in Europe. It also has the technology, the geographical and service diversification that will place them within the new global construction industry.

Despite the success of the Leighton Holdings, mergers and acquisitions are not always a guarantee for a long and successful life in the global industry, as the case of the Australian Concrete Construction illustrates. Taken over by Walter Bau-AG in 1995, it was placed in receivership in 2003 and the parent company followed it into liquidation in 2005. Although the German operation was later acquired by Bau Holding Strabag SE, Austria's largest construction group, there was little interest in the Australian operation, which was allowed to disappear, more or less without any trace.

Baulderstone Hornibrook is another typical example of an acquisition by what was to become a giant global operator. It was formed by the merger of AW Baulderstone Pty Ltd and MR Hornibrook Ltd and acquired in 1993 by Belfinger Berger AG, another German firm. While Bolderstone Hornibrook was one of Belfinger Berger's first major acquisitions, it certainly was not the last, and over the last 15 years, Bilfinger Berger AG has acquired or founded more than 25 major firms, including the Australian Abigroup, itself a result of several acquisitions of medium-sized contractors, including Enacon Limited, the Graham Evans Group, Robert Salzer Constructions, Hughes Bros Pty Limited and Simon Engineering.

Bilfinger Berger AG is now one of the top ten construction firms in Europe. It is not controlled by any individual major shareholder as the major parts of its shares are held by a wide range of institutional investors. As a result of its acquisitions, it is quite diversified. Its principal activity is structural and civil engineering including construction of railways, gas pipelines, bridges, roads, subways and tunnels, hydraulic engineering and offshore construction and the construction of prefabricated units, but also the construction and finances of residential and commercial property, water drainage, sewerage treatment and town planning. In terms of both the nature of the projects it undertakes and its size, it also belongs to the new construction industry.

While it is difficult to find accurate and meaningful figures, the projects listed as examples of the activities of these and other global firms demonstrate that what we have called a product, a project that combines several different services, is a significant factor in the development of the new industry. With the increased use of non-traditional procurement methods (different versions of design and build and operate) that combine design, construction and other services into a single entity, firms can appropriate the benefits of diversification, innovation, scale and research.

In construction, the methods used for tendering and procurement of projects are important determinants of the level and form of competition in the industry. The emergence of the one-stop supplier, the global firm, for large private and public projects, with their demand for capacity, technical capability and capital, has placed these projects out of reach of the conventional contractors (see also, Ezulike et al., 1997). The result is a 'two-tier' market. There is a market for both the traditional, fragmented, low-technology construction services supplied by the traditional construction industry and the high-technology products supplied by the new global industry.

Conclusion

In summary, the development of the new building industry, as we have envisaged it here, is facilitated by a changing environment. The most important forces behind the change are globalization of the market for construction and technological progress in communication technology. Mostly, however, it is internal to the firms, the result of attitude and preparations that enables firms to grasp the opportunities when they come and the vision to see where they leads them.

The new construction industry is an oligopolistic industry, with a small number of very large firms selling a differentiated product in a global market, competing through value for money. Because the product is differentiated, the firms will compete by technology, quality and design, as well as production costs. The difference between a differentiated product and an undifferentiated service which the rest of the construction industry provides, is far-reaching. Firms will have control over the design to ensure the use of product technologies and the market power to ensure the return on investments into the development of such technologies. By continually growing, the firms can exploit economies of scale and, by being diversified, they can not only provide more services for their clients but they can also apply new technologies over a range of new products, stimulating investments in R&D as they move into medium or high technology.

The result of this will be what is effectively a new construction industry that owes more to modern manufacturing than to the traditional, fragmented, small-scale and low-technology construction industry. The new industry will be a global oligopoly competing on value for money rather than price. It will be so different from the traditional industry that they will not compete with each other.

Notes

1 See Chapter 4 for a detailed treatise on issues of alternative methods of defining the construction sector.
2 The sources for this section are, unless otherwise stated, the websites of the respective firms.

References

Cole, G.D.H. (1920) *Chaos and Order in Industry.* Methuen and Co: London.

de Haan, J., Voordijk, H. and Joosten, G. (2002) Market strategies and core capabilities in the building industry. *Construction Management and Economics*, 20, 109–118.

Downie, J. (1958) *The Competitive Process.* Duckworth: London.

Egan, J. (1998) *Rethinking Construction.* Department of the Environment, Transport and the Regions: London.

Ezulike, E., Perry, J. and Hawwash, K. (1997) The barriers to entry into the PFI market. *Engineering, Construction and Architectural Management*, 4(3), 179–193

Hawk, D. (1992) *Forming a New Industry: International Building Production.* Swedish Council for Building Research: Solna, Svensk Byggtjanst.

Hawk, D. (2006) Conditions of success: a platform for international construction development. *Construction Management and Economics*, 24, 735–742.

Latham, M. (1994) *Constructing The Team.* Final Report of the Government/Industry Review of Procurement and Contractual Arrangements in the UK Construction Industry. HMSO: London.

Marris, R. (1964) *The Economic Theory of 'Managerial' Capitalism.* MacMillan: London.

Marris, R. (1972) Why economics needs a theory of the firm. *Economic Journal*, 82, Supplement.

Penrose, E. (1959) *The Theory of the Growth of the Firm.* Blackwell: Oxford.

Runeson, G. (2000) *Building Economics.* Deakin University Press: Geelong, Australia

Schumpeter, J. (1942) *Capitalism, Socialism and Democracy.* Harper: New York.

Seaden, G., Guolla, M., Doutriaux, J. and Nash, J. (2003) Strategic decisions and innovation in construction firms, *Construction Management and Economics*, 21, 603–612.

Chapter 11

The impact of reverse knowledge transfer on competitiveness

The case of Turkish contractors

M. Talat Birgonul, Irem Dikmen and Beliz Ozorhon

Introduction

The success of firms from developing economies attracts the attention of researchers in almost all sectors. Multinational firms from developing countries are shaking up entire industries and changing the rules of global competition (Business Week, 2006). The construction industry is not an exception to this trend. Although, construction companies from advanced industrialized countries (AIC) still play an important role in the global construction market as providers of construction services that require high technology and know-how, the global construction market is far from restricted to AICs (Bon, 1992). However, the theories in the international business literature are generally developed for the firms from AICs, which can hardly explain the current trends in the global market. In this chapter, the prevailing theories will be revisited and their validity for emerging economy multinationals that are involved in international construction will be discussed.

Theoretical background

One of the pioneer theories related to multinational enterprises is the 'eclectic theory of multinational advantage' developed by Dunning (1981, 1988). It is based on three types of advantages: advantage derived from extending resources such as brands, technology etc. (ownership advantage – O); advantage due to integration of resources worldwide with different costs in different locations (location advantage – L); and advantages derived through internalization of activities that would otherwise be dispersed between various firms (internalization advantage – I). In the 1990s, the OLI framework was revised to accommodate developments such as international partnerships and successful firms from emerging economies within the global arena. Mathews (2006) argues that the OLI framework and its revised versions are primarily based on transfer of superior resources abroad and, although they can explain why the first entrants to the global market (firms from AICs) are successful in the global market, they can hardly explain the competitive advantage of firms from developing countries.

The resource-based view sees organizations as bundles of resources combined with capabilities and core competencies that are primary determinants of performance (Hofer and Schendel, 1978; Wernefelt, 1984; Prahalad and Hamel, 1990; Barney, 1991). Based on this view, Aulakh (2007) explores the role of critical resources for success in the global market and pinpoints a debate between international business researchers about 'asset seeking' and 'asset exploitation' motivations behind internationalization. The question to be answered is whether ownership advantages and firm-specific capabilities precede or succeed international activity (Dunning, 2006; Mathews, 2006). Dunning (1988, 2006) explains internationalization as a process of transferring a firm's knowledge across borders. He argues that firms try to exploit their existing resource stock (ownership advantages) by internationalization. Thus, sources of competitive advantage for global firms are conceived to be the resources from the home market. On the other hand, Mathews (2006) puts forward the view that skills and knowledge gained as a result of international activity can also be a source of competitive advantage. Inward internationalization can act as a starting point to internationalization, as knowledge gained from foreign investors can be used to upgrade current skills and help to improve initial, usually technological, weaknesses. Likewise, outward internationalization may lead to improvement of these skills and lessons learnt as a result of international activity may result in further sources of competitive advantage. It is clear that, for the first entrants to the global market (incumbents as classified by Mathews, 2006), which are the firms from AICs, 'asset exploitation' may be the dominant motivator. However, 'asset seeking' behaviour better reflects the performance of firms from emerging economies (latecomers or newcomers as defined by Mathews, 2006).

Another popular theory which explains the success of firms in international markets is Porter's diamond framework (1998). The major idea behind Porter's diamond framework is that nations succeed in industries where their home base advantages are valuable in other nations. The main question is 'Why does a nation achieve international success in a particular industry?' According to Porter (1998), the answer lies in four broad attributes of a nation that shape the environment in which local firms compete, which are termed 'determinants of national competitive advantage'. These are factor conditions, demand conditions, related and supporting industries, firm strategy, structure and rivalry. Two additional variables can also influence the national system: chance and government. Chance events are outside the control of firms and even government. The diamond framework gives firms an insight into how to set strategy in order to become more effective against international competitors. Porter is not free from criticisms (Stopford and Strange, 1991; Dunning, 1993; Cho, 1994; Moon et al., 1998; Ofori, 2003; Ericsson et al., 2005). Dunning (1993) claims that the importance of multinational enterprises is underestimated in the diamond framework. Also, strategies are products of a learning process. Thus, companies learn from their international activities as well as their home market conditions. In order to eliminate this shortcoming, the multinational activity is defined as the third exogenous variable and the Dunning–Porter framework is formulated. Similarly, a 'generalized double diamond model' has been proposed by Moon et al. (1998) to incorporate global activity into the framework.

Mathews (2006) argues that in order to make sense of emerging economy multinationals' success in the global economy, an alternative framework is necessary. According to him, these firms' international expansion is driven by resource linkage, leverage and learning (LLL). In order to explain the concept of 'resource linkage', he argues that the firms from emerging economies are focused, not on their own advantages, but on the advantages that can be acquired externally, thus, global orientation becomes a source of advantage. Joint ventures and other forms of collaborative partnership are seen as principal vehicles for reducing risks and gaining competitive advantage. After firms access resources through linkage with external firms, ways are sought to leverage resources. Repeated processes of linkage and leverage may result in organizational learning to perform operations more effectively. The LLL framework may explain how initial weaknesses may be turned into core competencies with organizational learning as a result of repeated resource linkage and leverage and, finally, how the firms with limited resources can succeed in the global market.

In this chapter, the authors argue that 'learning from international activities' is the key concept that explains the success of global players coming from emerging economies and needs to be explored in more detail. Regarding the international activity of construction companies from emerging economies, research should be channelled to a learning theme rather than examination of the national sources of comparative advantage and ownership advantages only.

Review of studies on international construction business

Figure 11.1 depicts the major research routes in the field of international construction. The flow from 'home country to host country' (solid lines to the right), the links between the 'home country and the firm' as well as the 'firm and the host country' constitute the most popular research themes.

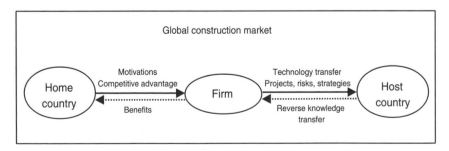

Figure 11.1 Research routes in the international construction literature.

Table 11.1 summarizes the research questions related with each topic depicted in Figure 11.1 and gives examples from previous studies in the international construction literature. The impact of the home country on the success of a firm (national comparative advantage) and sources of competitive advantage are among the topics that received the highest attention from researchers working in the field of international construction. Moreover, it is clear that the impacts of foreign investors on the home country (impact of inward internationalization) such as technology transfer and economic impacts have long been emphasized. However, the impact of international activity of domestic firms on the home market (impact of outward internationalization) has not been discussed extensively yet. It is argued that the dashed lines shown in Figure 11.1 constitute the research topics that are not widely covered in the international construction literature. These links demonstrate the impact of knowledge gained by international construction companies on the home country, which is a critical issue especially for firms of emerging economies. The link between the host country and the firm (as shown by a dashed line in Figure 11.1) demonstrates the extent of 'reverse knowledge transfer' that needs to be investigated in detail. This knowledge can be utilized to increase competitiveness in the global construction market, to increase competitive scope and may also have a major impact on the home country leading to major refinements in the home construction market. In this chapter, the impact of reverse knowledge transfer on the competitiveness of international firms in the global market is discussed rather than its impacts on the domestic market by referring to experiences of Turkish contractors.

Learning from international business: reverse knowledge transfer

Learning as a result of international activities is a popular research topic within the most recent international business literature (Ambos et al., 2006; Anh et al., 2006; Gugler and Brunner, 2007). Gugler and Brunner (2007) mention the benefits such as technology transfer, upgrading of human capital and dynamic upgrading of economy due to inward and outward internationalization. They also argue that there is a strong link between benefits of internationalization and national competitiveness. Learning from foreign partners and subsidiaries, which is a research theme that rests on knowledge-based theory (Grant, 1996; Nonaka, 1994; Nonaka and Takeuchi, 1995), is explored in various studies (Hakanson and Nobel, 2000, 2001; Foss and Pedersen, 2002; Zhou and Frost, 2003; Ambos et al., 2006).

It is believed that how the construction companies leverage their resources by reverse knowledge transfer needs to be examined in order to draw the overall picture of the internationalization process and to understand its real impacts. The impacts of reverse knowledge transfer also depend on the absorptive capacity of firms as well as absorptive capacity of the home market. The concept of

Table 11.1 Previous studies conducted in the field of international construction

Topic	Research questions	Previous studies conducted in the field of international construction
Home country	What are the government incentives to facilitate internationalization?	Raftery et al., 1998; Dulaimi et al., 2004
	What are the sources of national comparative advantage?	Seymour, 1987; Ofori, 1994; Oz, 2001; Cuervo and Pheng, 2003; Pheng and Hongbin, 2003, 2004
Home country → firm	Why do firms internationalize?	Gunhan and Arditi, 2005a
	What are the sources of competitive advantage?	Arditi and Gutierrez, 1991; Crosthwaite, 1998; Momaya and Selby, 1998; Pheng et al., 2004
Firm → host country	What are the critical success factors for winning and managing international projects?	Mawhinney, 2001; Howes and Tah, 2003; Gunhan and Arditi, 2005b
	What are the critical success factors for partnering?	Bing and Tiong, 1999; Mohamed, 2003; Gale and Luo, 2004; Chan et al., 2004, 2006; Ozorhon et al., 2007
	What are the risks of international projects?	Ashley and Bonner, 1987; Hastak and Shaked, 2000; Han and Diekmann, 2001; Dikmen and Birgonul, 2006; Kapila and Hendrickson, 2001
	What are the impacts of foreign investors on the country? (economic, technology transfer, reputation, etc.)	Ofori, 1990, 1994, 1996, 2007; Simkoko, 1992; Carrillo, 1994, 1996; Lopes, 1998; Lopes et al., 2002; Tan, 2002; Sexton and Barrett, 2004; Ruddock and Lopes, 2006; Ganesan and Kelsey, 2006, 2007
Host country	What are the market conditions prevailing in the host country? (volume of works, level of competition, risks, etc.)	Jaselskis and Talukhaba, 1998; Birgonul and Dikmen, 2001; Aleshin, 2001; Zarkada-Fraser and Fraser, 2002
Global construction market/environment	What are the patterns of international trade?	Hillebrandt, 2000; Bon, 1992, 1996, 1997; Bon and Crosthwaite, 2000, 2001; Ruddock, 2002; Chen and Messner, 2003; Howes and Tah, 2003
	What are the impacts of organizations (e.g. World Trade Organisation) on international trade?	Cuervo, 2003; Xu et al., 2005

absorptive capacity was first introduced by Cohen and Levinthal (1990) and defined as 'the ability to recognise the value of new information, assimilate it and apply it to create new knowledge and capabilities' (Anh et al., 2006). Teece et al. (1997) argue that resources can be improved, upgraded, can take an organization to another level and help to develop 'dynamic capabilities'. According to them, the expression *dynamic* reflects 'the capability of a firm to renew competencies so as to achieve congruence with the changing business environment' and the expression *capability* stresses 'the key role of strategic management in adapting, integrating and reconfiguring internal and external organisational skills, resources and functional competencies to match the requirements of a changing environment'. Dynamic capabilities may lead to sustainable competitive advantage especially in markets where the rules of competition change frequently. Reverse knowledge transfer by construction companies in international markets may help development of dynamic capabilities and lead to sustainable competitive advantage in the global market.

A conceptual model is developed in order to explore the impact of reverse knowledge transfer on the competitiveness of construction companies as shown in Figure 11.2. The model embraces the influences of absorptive capacity and competitive scope of a company on the development of dynamic capabilities and, finally, on the level of international competitive advantage. The propositions about the interrelations between the constructs are shown on the model and will be discussed by referring to the experiences of Turkish contractors working abroad.

Research methodology

Based on the conceptual model depicted in Figure 11.2, interviews were held with experts to explore the reverse knowledge transfer mechanism of Turkish

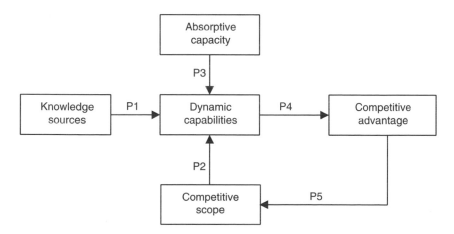

Figure 11.2 Conceptual model of reverse knowledge transfer.

contractors. The aim was to investigate the most important knowledge sources and the dynamic capabilities developed by Turkish contractors as a result of lessons learnt from international markets. Consequently, effective strategies may be devised to increase the rate of reverse knowledge transfer, absorptive capacity of firms and maximize the positive impact of international experience on the competitiveness of firms.

The experts were selected from the key parties that have different roles and potentially different viewpoints about the international activities of Turkish contractors. Interviews were conducted with four experts – two experts from each organization – representing the Turkish Contractors Association (TCA) and Coordination Board for International Contracting and Consulting Services (CBICCS). Three experts from three construction companies which are highly experienced in the international construction business were also interviewed.

TCA is an independent, non-profit professional organization whose members are the leading construction companies that have completed approximately 90% of international activities accomplished by Turkish contractors. Major objectives of TCA are to increase the competitiveness of its members in the national and international markets, to guide the government agencies and private bodies on legal, economic and technical issues, to encourage cooperation and mutual support among its members, and to raise public awareness on industry-related issues (TCA, 2007). CBICCS is a governmental body that addresses the problems of Turkish contractors working abroad. It has been established to coordinate the efforts of different stakeholders of the industry in an integral approach, develop strategies to encourage internationalization of Turkish contractors, and to ensure the benefits derived from this process.

Internationalisation of Turkish contractors

The construction sector is one of the leading sectors in Turkey that contributes to the national economy and employment due to its linkages with over 200 supplementary sectors. It is regarded as the locomotive sector since it contributes significantly to employment and gross national product (GNP). Construction investment corresponds to 50 per cent of total investment and has a direct share of 5 per cent and an indirect share of 33 per cent in GNP distribution (Akguloglu, 2007). Turkey's share in the total world construction volume is around 3 per cent. Based on low-cost, rapid construction ability, qualified human resources, technical expertise and experience, Turkish contractors achieved an important competitive edge in international markets. As of 2006, the total volume of the works carried out by Turkish contractors abroad has reached US$5 billion within 63 countries and the number of completed projects is over 3,600. It is estimated that this figure will be US$100 billion by the end of 2007 (TCA, 2007). Currently, there are 22 Turkish contractors among the leading 225 international construction companies in the Engineering News Record list (ENR, 2007).

Overseas experiences of the Turkish contractors started at the beginning of 1970s when they first exported their services to Libya. Turkish companies expanded their businesses in Arab countries such as Iraq, Jordan, Saudi Arabia, Kuwait, the United Arab Emirates, Yemen and Iran. Since the beginning of the 1980s, former Soviet Union countries have become the target markets. In the 1990s, due to the political uncertainties and economic depression in the Middle Eastern and North African countries, Turkish contractors have focused on the Commonwealth of Independent States, Eastern Europe and Asian countries. In this period, important projects were completed in the Russian Federation, Ukraine, the Caucasus, the Central Asian Republics, Germany, Pakistan and the Far East. As of today, Turkish contractors operating in 63 countries across four continents have achieved a competitive edge providing services compatible with international contracting standards in terms of financial, administrative and technological aspects. So far, they have mainly concentrated their oper ations in the Eurasian region and recently they have been seeking opportunities in the African market. Latin America and Southeast Asia are also observed to be attractive markets. Meanwhile, it is foreseen that oil-rich countries of the Middle East, including Iraq, will be maintained as alternative targets, provided that political stability is secured in that region (TCA, 2007).

Propositions

There are five major propositions, the validity of which were investigated with the experts participating in this study:

P1: The formation of dynamic capabilities depends on knowledge acquisition from different sources in the international market

The major knowledge sources for Turkish companies, as mentioned during the interviews, are joint venture partner(s), client and third parties. Knowledge gained from the overseas projects can be categorized as procedural (know-how) and declarative (know-what) knowledge. The facts related to the international projects, market conditions and contractual issues may be regarded as declarative, whereas the technical and managerial skills are procedural knowledge. FIDIC conditions of contract is an example of declarative knowledge that the Turkish contractors acquired in international markets. Familiarity with international laws, regulations and standard form of contracts is the initial step of effective contract management, which further may become a source of competitive advantage in global markets.

In accordance with the resource linkage concept as put forward by Mathews (2006), the major sources of Turkish contractors were not exceptional resources or capabilities. However, they established effective partnerships to leverage their limited resources. As all of the experts agreed, establishing joint ventures is an effective solution for construction firms to enhance their technological know-how and to acquire new skills. It was interesting that Turkish contractors denote 'managerial' rather than 'technical and technological' issues among the most important types of knowledge learnt from their partners. Turkish contractors recognized the importance of risk management and claim management topics. They also learnt how to set up effective project management systems and carry out systematic project management tasks. The importance of communication in cross-cultural teams became apparent while conducting business abroad. Discussion sessions indicated that contractors learn a considerable amount of declarative knowledge from owners' representatives and third parties. Working with experienced consultants and client's representatives, they increased their awareness of the importance of data collection from site and documentation systems, contract management, quality assurance and occupational health and safety issues. In the international projects, Turkish contractors were also exposed to a different approach in which consultants, contractors and clients have to collaborate closely, which was a rather new concept for the Turkish construction industry. This kind of experience helped them focus on 'collaboration' and 'negotiation' rather than 'competition' between project participants.

P2: The formation of dynamic capabilities depends on the competitive scope of firms

As explained in the previous section, internationalization of Turkish contracting services began in the early 1970s through the Turkish–Libyan Joint Economic Cooperation Protocol that provided valuable overseas experiences (Oz, 2001). Within the period 1970–2000, Turkish contractors undertook projects mainly in Libya, Russian Federation, Saudi Arabia, Iraq, Kazakhstan, Turkmenistan and Azerbaijan. The first ranking activity undertaken by Turkish contractors during those years was housing projects. The sources of competitive advantage in those markets have been low labour cost, rapid construction, and geographical, religious and cultural proximity to the host countries. Thus, Turkish contractors needed to devise strategies for low-cost and rapid construction issues. For example, using high-quality, low-price materials exported from Turkey and employing Turkish labour in the host country created, for them, a low-cost advantage. Moreover, as most of the projects were carried out in high-risk countries, they had to develop the capability of pricing country risk realistically and estimating proper contingency values. As the competitive scope changes, the required capabilities to sustain or improve competitiveness also change. The validity of this issue for the Turkish contractors is discussed under proposition P5.

P3: The formation of dynamic capabilities depends on the absorptive capacity of firms

Reverse knowledge transfer may bring its benefits if the absorptive capability of firms is high. Declarative and procedural knowledge gained in international projects can be converted to dynamic capabilities as far as the absorptive capacity of firms allows. Absorptive capacity is partly determined by the strategic perspective of a company and partly by the home market conditions. Regulations, national policies and demand in the domestic market are major enablers/barriers in developing countries to improve domestic capacity, productivity and quality in materials, equipment and labour, procurement and collaboration practices, organizational, managerial, design and construction skills (Ganesan and Kelsey, 2007). If the domestic market does not apply international standards, leading to a low level of absorptive capacity, it becomes inevitable that the companies leave behind what is learnt from international markets, which is partially the case for the Turkish construction industry. The main source of competitive advantage of contractors in the Turkish construction market is low cost. Since the domestic market conditions are not competitive in terms of productivity, quality, technology, health and safety and environmental issues, companies do not need to develop capabilities which would enable them to transfer their knowledge and skills to domestic projects.

As well as the competitive rules prevailing in the domestic market, the strategic perspective of a company and existence of management systems for internalization of knowledge gained from various sources are important indicators of its absorptive capacity. If some mechanisms (such as post-project appraisals, writing learning histories etc.) and cultural background exist to eliminate the risk of organizational amnesia, the absorptive capacity is high and the knowledge gained from international markets becomes re-usable, creating a source of competitive advantage. Discussions with experts revealed the fact that, although the domestic market conditions (local procurement law, type of demand, etc.) partially hinder the transfer of knowledge gained from international markets to the domestic market, companies may concentrate on the international market rather than the domestic market and, in that case, they have to increase their absorptive capacity by establishing effective knowledge management systems within their company. The strategies identified by the experts to increase absorptive capacity are establishing proper information systems so that the knowledge gained can be distributed within the company, transforming the culture of the company to 'an organizational learning culture' and launching a reward mechanism to motivate people to share their knowledge.

P4: Dynamic capabilities create competitive advantage

As mentioned earlier, the major sources of competitive advantage of Turkish contractors operating in developing markets such as Central Asia, Middle East and North Africa are geographical, religious and cultural proximity, rapid construction under adverse conditions and inexpensive workforce. However, in those markets, competition is becoming higher and more severe due to the high number of contractors from other emerging economies (e.g. Korean and Chinese) that also have inexpensive labour and governmental support, resulting in lower profits. Therefore, Turkish companies had to develop new capabilities and competencies by transferring knowledge gained from their foreign partners in international projects. In this respect, reverse knowledge transfer contributed considerably to the restructuring of the companies; technological, managerial and multinational communication skills were improved and awareness on quality, health and safety issues was increased. The differentiation skills gained, based on the international experience (resource linkage, leverage and learning as mentioned by Mathews (2006)) opened the door of new markets and increased their competitiveness.

P5: The competitive advantage attained by firms may lead to a change in competitive scope

Using the knowledge gathered from their partners and other parties, Turkish contractors had the chance to enter new markets and realize technically and managerially more complex projects in the international arena. Those companies that were able to leverage their resources and capabilities increased their competitive scope. After the year 2000, regarding the competitive scope of Turkish contractors, there has been a considerable shift from housing to infrastructure and industrial projects that require advanced technological and project management skills as well as experience. Currently, Turkish contractors are seeking projects in Latin America, Africa and Southeastern Asia as well as in the Balkans and the Middle East, meaning that low-cost, rapid construction may not be sufficient to create competitive advantage. They have to differentiate their services by developing necessary competencies such as project development ability, utilization of IT to support management and construction processes, establishing long-term strategic alliances as well as short-term partnerships and development of integrated management systems (quality, environmental impact, health and safety issues). The knowledge gained from various sources in international markets and experience will be used to create required sources of competitive advantage in new markets.

It was emphasized during interviews that Turkish contractors are also competing based on positive company image and reputation (developed as a result of previous activities in the international markets). They established strong relations in international markets and developed a positive reputation which helps them increase their competitive scope. Moreover, as well as contracting jobs in new markets, they have extensively diversified their business by investing in other sectors such as real estate development, hotel and airport management.

They are not only giving contracting services but also acting as developers, material suppliers and operators in major projects.

Figure 11.2 shows a feedback loop between dynamic capabilities, competitiveness and competitive scope. Competitiveness is increased by developing dynamic capabilities which, in turn, may result in a change in competitive scope that requires new dynamic capabilities. Turkish experience has demonstrated that companies from emerging economies may start from a point where they have limited resources and capabilities, but they may increase their competitiveness in the long run with repeated cycles of linkage, leverage and learning.

Conclusions

The aim of this research was to examine the reverse knowledge transfer of Turkish contractors working abroad and propose some strategies to increase their competitiveness in international markets. Interviews carried out with experts revealed the fact that Turkish contractors initially entered into international markets by using their cultural and religious similarities with the host countries and had serious resource limitations. Resource linkage was achieved mainly through partnering; resource leverage was facilitated by learning from a number of sources; and finally, they developed dynamic capabilities which in turn resulted in a wider competitive scope. Contract, claim, risk and cross-cultural management skills as well as technical skills are among the major strengths developed by Turkish contractors through partnering. Moreover, overseas projects contributed significantly to their awareness on health and safety issues, quality management, environmental impact and its assessment.

Although international contracting offers several opportunities in terms of enriching organizational assets, broader benefits can be achieved through proper strategies and mechanisms at company and governmental levels. In order to increase national competitiveness and to maximize the positive impacts of internationalization on the domestic market, reverse knowledge transfer should be accelerated by enhancing both the companies' and the domestic market's absorptive capabilities.

The organizational structure and culture of the companies are highly influential in increasing their organizational learning ability. Organizational learning ability can be improved by devising necessary systems for knowledge acquisition, storage, sharing and retrieving as well as establishing an appropriate cultural background. It is suggested that companies should focus more on internalizing knowledge and skills acquired from international activities to sustain competitive advantage. Collaboration of contractors with non-governmental bodies in Turkey such as TCA, Turkish Employers' Association of Construction Industries, etc. may also be helpful in designing and implementing practical systems that would facilitate the knowledge-integration process. Mechanisms employed by the government should be targeted to bridge the gap between the rules and regulations prevailing in the Turkish construction industry and international standards, so that the absorptive capacity of the domestic construction market can be increased. A governmental support programme has been initiated to support the internationalization of Turkish contractors, to identify and solve problems of the industry and to coordinate the efforts of institutions contributing to international contracting services. This programme should also deal with the issue of reverse knowledge transfer and institutionalization of lessons learnt by Turkish contractors from international markets. Also non-governmental bodies, professional societies and associations are indispensable parts of the industry that can assist governmental organizations about how to transfer lessons learnt in international markets and revise the domestic market conditions to take full advantage of dynamic capabilities developed by the Turkish contractors as a result of international activity.

As a final remark, Turkish contractors' experience in international markets shows that, although the initial impetus for internationalization of services was national comparative advantage, the sustainability of success in international markets can be explained by the resource linkage, leverage and learning framework put forward by Mathews (2006). The reverse knowledge transfer which is synchronized with the learning process should be examined in detail so that the interrelations between the dynamic capabilities, competitive scope, absorptive capacity and competitiveness can be revealed. Consequently, strategies and policies can be developed to achieve the required level of international competitiveness in target markets and to maximize benefits for the domestic market.

References

Akguloglu, K. (2007) *Yurtdışı Müteahhitlik ve Teknik Müşavirlik Hizmetleri*, available online at http://www.korayakguloglu.net/dosyalar/20032007084247YDMH1.2007.pdf

Aleshin, A. (2001) Risk management of international projects in Russia. *International Journal of Project Management*, 19(4), 207–222.

Ambos, T.C., Ambos, B. and Schlegelmilch, B.B. (2006) Learning from foreign subsidiaries: an empirical investigation of headquarters' benefits from reverse knowledge transfers. *International Business Review*, 15, 294–312.

Anh, P.T.T., Baughn, C.C., Hang, N.T.M. and Neupert, K.E. (2006) Knowledge acquisition from foreign parents in international joint ventures: an empirical study in Vietnam. *International Business Review*, 15, 463–487.

Arditi, D. and Gutierrez, A. (1991) Factors affecting US contractors' performance overseas. *ASCE Journal of Construction Engineering and Management*, 117(1), 27–46.

Ashley, D.B. and Bonner, J.J. (1987) Political risks in international construction. *Journal of Construction Engineering and Management*, 113(3), 447–467.

Aulakh, P.S. (2007) Emerging multinationals from developing economies: motivations, paths and performance. *Journal of International Management*, 13(3), 235–240.

Barney, J. (1991) Firm resources and sustained competitive advantage. *Journal of Management*, 17, 99–120.

Bing, L. and Tiong, R.L.K. (1999) Risk management model for international construction joint ventures. *ASCE Journal of Construction Engineering and Management*, 125(5), 377–384.

Birgonul, M.T. and Dikmen, I. (2001) Risks borne by foreign contractors doing business in Turkey. In: Akintoye, A. (ed) *Proceedings of ARCOM (Association of Researchers in Construction Management)*, University of Salford, 5–7 September, Manchester, UK, Vol. 2, pp. 845–853.

Bon, R. (1992) The future of international construction: secular patterns of growth and decline. *Habitat International*, 16(3), 119–128.

Bon, R. (1996) Whither global construction? Some results of the ECERU opinion survey, 1993–95. *Building Research and Information*, 24(2), 81–85.

Bon, R. (1997) The future of international construction: some survey results, 1993–96. *Building Research and Information*, 25(3), 137–141.

Bon, R. and Crosthwaite, D. (2000) *The Future of International Construction: Collected Papers in Input–Output Modelling and Applications*. Thomas Telford: London.

Bon, R. and Crosthwaite, D. (2001) The future of international construction: some results of 1992–1999 surveys. *Building Research and Information*, 29(3), 242–247.

Business Week (2006) Emerging giants. *Business Week*, Cover Story, July 31.

Carrillo, P. (1994) Technology transfer: a survey of international construction companies. *Construction Management and Economics*, 12, 45–54.

Carrillo, P. (1996) Technology transfer on joint-venture projects in developing countries. *Construction Management and Economics*, 14(1), 45–54.

Chan, A.P.C., Chan, D.W.M., Chiang, Y.H. et al. (2004) Exploring critical success factors for partnering in construction projects. *ASCE Journal of Construction Engineering and Management*, 130(2), 188–198.

Chan, A.P.C., Chan, D.W.M., Fan, L.C.N. et al. (2006) Partnering for construction excellence – a reality or myth? *Building and Environment*, 41, 1924–1933.

Chen, C. and Messner, J.I. (2003) An investigation of entry-remain-exit patterns in international construction markets. In: Ofori, G. and Ling, F.Y.Y. (eds) *Proceedings of the Joint International Symposium of CIB Working Commissions*, Singapore, 22–24 October, National University of Singapore, Singapore, Vol. 2, pp. 329–340.

Cho, D.S. (1994) A dynamic approach to international competitiveness: the case of Korea. *Journal of Far Eastern Business*, 1(1), 17–36.

Cohen, W. and Levinthal, D. (1990) Absorptive capacity: a new perspective on learning and innovation. *Administrative Science Quarterly*, 35, 128–152.

Crosthwaite, D. (1998) The internationalization of British construction companies 1990–96: an empirical analysis. *Construction Management and Economics*, 16, 389–395.

Cuervo, J.C. (2003) The WTO system: impact on construction services. In Ofori, G. and Ling, F.Y.Y. (eds) *Proceedings of the Joint International Symposium of CIB Working Commissions*, Singapore, 22–24 October, National University of Singapore, Singapore, Vol. 2, pp. 341–351.

Cuervo, J.C. and Pheng, L.S. (2003) Ownership advantages/disadvantages of Singapore transnational construction corporations. *Construction Management and Economics*, 21(1), 81–94.

Dikmen, I. and Birgonul, M.T. (2006) A review of international construction research: Ranko Bon's contribution. *Construction Management and Economics*, 24(7), 725–733.

Dulaimi, M.F., Ling, F.Y.Y. and Ofori, G. (2004) Engines for change in Singapore's construction industry: an industry view of Singapore's Construction 21 report. *Building and Environment*, 39, 699–711.

Dunning, J.H. (1981) *International Production and the Multinational Enterprise*. Allen and Unwin: London.

Dunning, J.H. (1988) The eclectic paradigm of international production: a restatement and some possible extensions. *Journal of International Business Studies*, 79, 1–31.

Dunning, J.H. (1993) *Multinational enterprises and the global economy*. Addison-Wesley: Wokingham.

Dunning, J.H. (2006) Towards a new paradigm of development: implications for the determinants of international business. *Transnational Corporations*, 15(1), 173–227.

Engineering News Records (ENR) (2007) The top 225 international contractors. *Engineering News Records*, August 20, 38–44.

Ericsson, S., Henricsson, P. and Jewell, C. (2005) Understanding construction industry competitiveness: the introduction of the hexagon framework. *Proceedings of CIB Conference on Combining Forces*, Helsinki, Finland, Vol. 2, pp. 186–202.

Foss, N.J. and Pedersen, T. (2002) Transferring knowledge in MNCs: The role of sources of subsidiary knowledge in organizational context. *Journal of International Management*, 8(1), 49–67.

Gale, A. and Luo, J. (2004) Factors affecting construction joint ventures in China. *International Journal of Project Management*, 22(1), 33–42.

Ganesan, S. and Kelsey, J.M. (2006) Technology transfer: international collaboration in Sri Lanka. *Construction Management and Economics*, 24, 743–753.

Ganesan, S. and Kelsey, J.M. (2007) Construction in Shanghai: issues in technology transfer', *CME 25 Conference, Construction Management and Economics 'Past, Present and Future'*, 16–18 July, University of Reading, UK.

Grant, R.M. (1996) Toward a knowledge-based theory of the firm. *Strategic Management Journal*, 17, 109–122.

Gugler, P. and Brunner, S. (2007) FDI effects on national competitiveness: a cluster approach. *International Advances in Economic Research*, 13(3), 268–284.

Gunhan, S. and Arditi, D. (2005a) Factors affecting international construction. *ASCE Journal of Construction Engineering and Management*, 131(3), 273–282.

Gunhan, S. and Arditi, D. (2005b) International expansion decision for construction companies. *ASCE Journal of Construction Engineering and Management*, 131(8), 928–937.

Hakanson, L. and Nobel, R. (2000) Technology characteristics and reverse knowledge transfer. *Management International Review*, 40(special issue), 29–48.

Hakanson, L. and Nobel, R. (2001) Organizational characteristics and reverse knowledge transfer. *Management International Review*, 41(4), 395–420.

Han, S. and Diekmann, J.E. (2001) Approaches for making risk-based go/no-go decision for international projects. *ASCE Journal of Construction Engineering and Management*, 127(4), 300–308.

Hastak, M. and Shaked, A. (2000) ICRAM-1: Model for international construction risk assessment. *ASCE Journal of Management in Engineering*, 16(1), 59–69.

Hillebrandt, P.M. (2000) *Economic Theory and the Construction Industry*, 3rd edition. Macmillan: Basingstoke.

Hofer, C.W and Schendel, D. (1978) *Strategy Formulation: Analytic Concepts*. West Publishing Company: St. Paul, MN.

Howes, R. and Tah, J.H.M. (2003) *Strategic Management Applied to International Construction*. Thomas Telford Publishing: London.

Jaselskis, E.J. and Talukhaba, A. (1998) Bidding considerations in developing countries. *ASCE Journal of Construction Engineering and Management*, 124(3), 185–193.

Kapila, P. and Hendrickson, C. (2001) Exchange rate risk management in international construction ventures. *ASCE Journal of Management in Engineering*, 17(4), 186–191.

Lopes, J. (1998) The construction industry and macroeconomy in Sub-Saharan Africa post 1970. *Construction Management and Economics*, 16, 637–649.

Lopes, J., Ruddock, L. and Ribeiro, F.L. (2002) Investment in construction and economic growth in developing countries. *Building Research and Information*, 30(3), 152–159.

Mathews, J.A. (2006) Dragon multinationals: new players in 21st century globalization. *Asia Pacific Journal of Management*, 23(1), 5–27.

Mawhinney, M. (2001) *International Construction*. Blackwell Science: London.

Mohamed, S. (2003) Performance in international construction joint ventures: modelling perspective. *ASCE Journal of Construction Engineering and Management*, 129(6), 619–626.

Momaya, K. and Selby, K. (1998) International competitiveness of the Canadian construction industry: a comparison with Japan and the United States. *Canadian Journal of Civil Engineering*, 25(4), 640–652.

Moon, C.H., Rugman, A.M. and Verbeke, A. (1998) A generalized double diamond approach to the global competitiveness of Korea and Singapore. *International Business Review*, 7, 135–150.

Nonaka, I. (1994) A dynamic theory of organizational knowledge creation. *Organization Science*, 5, 14–37.

Nonaka, I. and Takeuchi, H. (1995) *The Knowledge Creating Company: How Japanese Companies Create the Dynamics of Innovation*. Oxford University Press: London.

Ofori, G. (1990) *The Construction Industry: Aspects of its Economics and Management*. Singapore University Press: Singapore.

Ofori, G. (1994) Construction industry development: role of technology transfer. *Construction Management and Economics*, 12, 379–392.

Ofori, G. (1996) International contractors and structural changes in host country construction industries: case of Singapore. *Engineering, Construction and Architectural Management*, 3(4), 271–288.

Ofori, G. (2003) Frameworks for analysing international construction. *Construction Management and Economics*, 21(4), 379–391.

Ofori, G. (2007) Construction in developing countries. *Construction Management and Economics*, 25(1), 1–6.

Oz, O. (2001) Sources of competitive advantage of Turkish construction companies in international markets. *Construction Management and Economics*, 19, 135–144.

Ozorhon, B., Arditi, D., Dikmen, I. and Birgonul, M.T. (2007) Effect of host country and project conditions in international construction joint ventures. *International Journal of Project Management*, 25(8), 799–806.

Pheng, L.S. and Hongbin, J. (2003) Internationalization of Chinese construction enterprises. *Journal of Construction Engineering and Management*, 129(6), 589–598.

Pheng, L.S. and Hongbin, J. (2004) Estimation of international construction performance: analysis at the country level. *Construction Management and Economics*, 22(3), 277–289.

Pheng, L.S., Jiang, H. and Leong, C.H.Y. (2004) A comparative study of top British and

Chinese international contractors in the global market. *Construction Management and Economics*, **22**(7), 717–731.

Porter, M.E. (1998) *The Competitive Advantage of Nations*. Macmillan Business: London.

Prahalad, C.K. and Hamel, G. (1990) The core competence of the corporation. *Harvard Business Review*, **68**(3), 79–91.

Raftery, J., Pasadilla, B., Chiang, Y.H. et al. (1998) Globalization and construction industry development: implications of recent developments in the construction sector in Asia. *Construction Management and Economics*, **16**, 729–737.

Ruddock, L. (2002) Measuring the global construction industry: improving the quality of data. *Construction Management and Economics*, **20**(7), 553–556.

Ruddock, L. and Lopes, J. (2006) The construction sector and economic development: the 'Bon curve'. *Construction Management and Economics*, **24**(7), 717–723.

Sexton, M.G. and Barrett, P. (2004) The role of technology transfer in innovation within small construction firms. *Engineering, Construction and Architectural Management*, **11**(5), 342–348.

Seymour, H. (1987) *The Multinational Construction Industry*. Croom Helm: London.

Simkoko, E.E. (1992) Managing international construction projects for competence development within local firms. *International Journal of Project Management*, **10**(1), 12–22.

Stopford, J.M. and Strange, S. (1991) *Rival States, Rival Firms: Competition for World Market Shares*. Cambridge University Press: Cambridge.

Tan, W. (2002) Construction and economic development in selected LDCs: past, present and future. *Construction Management and Economics*, **20**(7), 593–599.

Turkish Contractors Association (TCA) (2007) *Turkish Contracting in the International Market*. Available online at http://www.tmb.org.tr/genel.php?ID=10.

Teece, D.J., Pisano, G. and Shuen, A. (1997) Dynamic capabilities and strategic management. *Strategic Management Journal*, **18**(7), 509–533.

Wernefelt, B. (1984) A resource-based view of the firm. *Strategic Management Journal*, **5**, 171–180.

Xu, T., Tiong, R.L.K., Chew, D.A.S. and Smith, N.J. (2005) Development model for competitive construction industry in the People's Republic of China. *ASCE Journal of Construction Engineering and Management*, **131**(7), 844–853.

Zarkada-Fraser, A. and Fraser, C. (2002) Risk perception by UK firms towards the Russian market. *International Journal of Project Management*, **20**(2), 99–105.

Zhou, C. and Frost, T. (2003) Centrifugal forces, R&D co-practice, and 'reverse knowledge flows' in multinational firms. *Paper presented at AIB annual meeting*, 5–8 July, Monterey, California.

Chapter 12

Market interdependencies between real estate, investment, development and construction

The Dutch experience

Jo P. Soeter, Philip W. Koppels and Peter De Jong

Introduction

Demand in the real estate market is driven by a combination of population growth, employment growth, economic growth and the interrelated growth of the stock of buildings. These traditional growth drivers are weakening in Western Europe. Nowadays investors, developers and construction firms need a strategy for dealing with the structural shift, conjunctural volatility and growing risks of operations on the real estate market and its different segments. Free market conditions under European constraints do not guarantee that the supply side of the construction market (labour/production capacity and design–build–finance–maintain know how and services) matches well with changing demand characteristics. The latter are influenced by building for a changing society and economy, for modernization of the stock of buildings by replacement and renovation and for sustainability.

Analysis and modelling of market development and market interdependencies provides for a better understanding of real estate in terms of economic context and past and future development.

The rise of the Dutch built environment

In the 1960s, investment in buildings grew very rapidly (Figure 12.1). Post-war reconstruction activity was insufficient to fulfil the need for buildings caused by a growing population and a growing economy. Nowadays, 40 years later, the building stock originating from the 1960s is an object of investment for renewal.

At the beginning of the twenty-first century, investment in dwellings is four times higher than that of the early 1960s. In the first investment peak in the early 1970s, large-scale, high-rise, social rented housing was overrepresented. Gradually, the number of new constructed dwellings decreased and investment per dwelling increased. Furthermore, the average project scale diminished and investment in upgrading of the existing stock became more important. The experience in the 1990s was that residential construction still grew while the non-residential sector diminished. Since 2003, the same trend can be observed again.

In the 1960s, non-residential building investment grew to the same extent as residential building investment. After a first peak in the early 1970s, non-residential building investment reveals a cyclical loop with an overall peak in 2001.

In the early 1980s, non-residential building investment diminished strongly. In the mean time, economic growth diminished and the nominal interest rate rose above 10 per cent. The Dutch economy suffered from an international economic crisis and from a financial crisis. The government budget deficit grew, which led to a growing government debt and an increased interest obligation, which created new deficits. Direct and indirect labour costs grew and became an obstacle for the growth of export of goods and services. Nevertheless, the balance of payments (exports minus imports) did not create further problems, because the Dutch economy maintained its export surplus from increased natural gas exports in combination with rising energy prices.

Since 1983, rapid growth of non-residential building has been part of an overall recovery of the Dutch economy. Better control and a strict limitation of the government deficit, debt and labour costs contributed to growing economic initiatives and capital formation by the market sector.

The quantitative link between changes in investment in non-residential

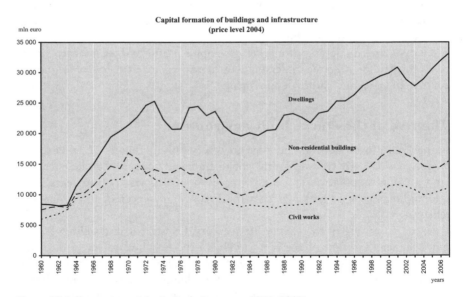

Figure 12.1 Formation of the built environment 1960–2007.

Source: Statistics Netherlands (CBS)/elaboration TU Delft – Building Economics.

buildings and changes in investment in machinery was weak and indirect in the 1990s. Between 1995 and 2001 office construction, in particular, was booming (Figure 12.2). This correlated with the growth of investment in ICT (computers, databases, software). After 2001, total fixed investment declined.

New construction is not a prerequisite for further technological development of the stock of buildings. The existing stock adapts to new ICT facilities as well. Furthermore, investment in machinery and ICT has a rather short turn-over time in comparison with investment in buildings. The replacement of, and the depreciation on, the stock of machinery and ICT is relatively high. Since 2004, investment in non-residential buildings has shown a weak recovery. Growth is stronger for machines and computers.

For developers and construction firms, the current real estate and construction market is dominated by a growing demand for residential development and construction. In the mean time, the non residential market, and especially office construction, is slowly recovering from an economic downturn. Thanks to the public sector (government, health services, education, public services) there was a weak recovery in 2006.

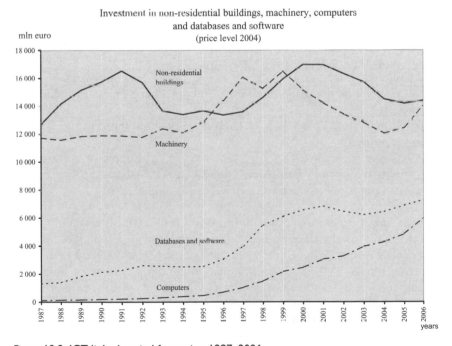

Figure 12.2 ICT linked capital formation 1987–2006.

Source: Statistics Netherlands (CBS)/elaboration TU Delft – Building Economics.

The overall peak of construction activities in 2001 created serious tension on the construction market, which was a seller's market in the period 1995–2001.

Interrelated markets

The *modern built environment* is provided by an industry which nowadays has to focus on final demand. The end-users have to pay rent and this determines the direct return on investment for the investors. The users can choose between various accommodation alternatives. However, the investors have to be aware of the value in use; in order to prevent vacancy, they have to choose their properties well. Within the supply chain the developers (and designers) have to identify the requirements and wishes of the known and unknown end-users.

The analysis of market adaptation, returns and risks has to distinguish between demand (D) and supply (S) on interrelated markets:

- Space market: $\Delta D_{stock} > = < \Delta S_{stock}$ ($> = <$ means demand is larger than, equal to or lower than supply). New users (tenants and owner-occupiers) demand expansion and renewal of the stock of buildings. Synonyms for space market are usage market and rental market. The overall space market is divided into submarkets like the office space market, the retail space market, the industrial space market etc.
- Property market: $\Delta D_{property} > = < \Delta S_{property}$. Investors and owner-occupiers demand new and renewed property, which is supplied by developers etc.
- Capital market: $D_{capital} > = < S_{capital}$. New owners and developers demand for property finance and development.
- Development market: $D_{development} > = < S_{development}$. New initiatives by developers etc. in anticipation of effective and potential demand for additional property.
- Land market: $D_{land} > = < S_{land}$. Land is an essential input in the development process.
- Construction market: $D_{construction} > = < S_{construction}$. Supply of construction capacity is limited by bottleneck inputs, such as the labour capacity and the capacity of supply industries (materials and components).

In the long run, economies are in transition from take-off to rapid growth and eventually to a mature economy. In rapidly growing economies, like the Netherlands in the 1960s, all the mentioned market relations tend to demand exceeding supply ($D > S$). New investment in buildings is indirect, limited by the bottleneck in the production capacity (e.g. labour shortage) or by shortage of capital. Seller's market conditions dominate all distinguished market relations. Reduced vacancy rates, rising prices and costs, guaranteed returns and low risks accompany the real estate development and financial management.

In mature economies, like most West European countries nowadays, with a mature stock of buildings, there is a higher chance that one or more market conditions change into oversupply (D < S). The growing vacancy of old buildings, undersupply of modern property for investment, oversupply of capital and bottlenecks in the production capacity may occur at the same time. Consequently, the return–risk profiles change.

Di Pasquale and Wheaton (1996) and Geltner et al. (2007) combine analysis of the space market, the asset market and the development and construction market in a four quadrant scheme modelling. In this chapter, a macro concept is chosen for the building market, which will be analysed in the chain space market, property market and development and construction market.

In Figure 12.3, the four axes and quadrants are defined as:

- *North axis:* represents the real annual payment in euros per unit, equivalent to costs of finance for owner-occupiers or net rent for the tenant.
- *East axis:* represents the gross and net investment in buildings (land excluded) in a certain year. Net investment is for expansion of the stock of buildings. Gross investment consists of net investment plus investment for renewal of the stock of buildings.
- *NE quadrant (space market):* illustrates demand functions for net and gross investment in buildings as a function of annual payment. Shifts in

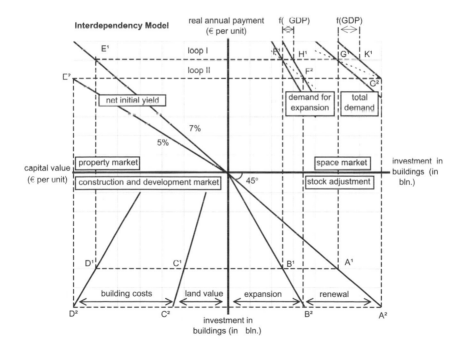

Figure 12.3 Interdependency model.

the demand function are dependent on changes in the gross domestic product.

- *West axis:* represents capital value of property per unit.
- *NW quadrant (property market):* expresses the capitalization of the annual payments to capital value. The relationship between annual payment and real estate value is visualised as a radius with a slope that is dependent on the net initial yield (NIY). The NIY-% here is equivalent to a 30-year real annuity in case of debt finance of property.
- *South axis:* represents the gross investment in buildings equal to east axis (in billion euros).
- *SW quadrant (development and construction market):* illustrates the supply cost function for investment in buildings. This function is here derived by combining the capital value on the west axis with the investment level (building activity) on the south axis. The capital value of property has to be divided into land value and building value which are normally analysed and modelled as 'building costs' and 'residual land value'.
- *SE quadrant:* in this quadrant, the 45° line serves as axis transformator. The south axis is equal to the east axis. Gross investment is split into net investment and investment for renewal of the stock of buildings. An exogenous analysis is employed to estimate the share of net investment of the gross investment.

To conclude, the demand functions in the NE quadrant are based on the investment levels in the SE quadrant, which are combined with the annual payment levels on the north axis. The shifts of the demand functions are based on changes in the GDP.

Loop I reflects a high real annual payment and a high NIY. In loop II, both values are lower and that leads to a rise in capital value. The Dutch experience in the period 1996–2001 is that net rent does not decline when real interest declines; there is still an increase. Dutch rent levels show inelasticity for diminishing NIY and NIY reacts on diminishing real interest with a lag.

Bak (2007) presents data for a medium-sized office project development. Published real rents rose from €145 in 1996 to €160 in 2006 (2004 prices). NIYs in the market were almost 7 per cent in 1996, 6.5 per cent in 2001 and 5.5 per cent in 2006. The 1996 and 2001 yields lagged behind the diminishing real interest. In 2006, the opposite occurred. NIYs decreased, while the real interest rate was higher than in 2001. In Table 12.1, data are shown on gross rent, capital value, net rent, net initial yield, building costs and residual land value.

In the late 1990s, a high demand caused the initial rents for newly constructed offices to rise at a higher rate than general inflation. NIYs showed a fractional decline of 0.5 per cent.

After 2001, published real rents did not drop. Nevertheless effective rents declined, since lessors offered the tenants lease-free periods or other incentives,

Table 12.1 Medium office project development (Euros (2004)) (LFA = lettable floor area; GFA = gross floor area; building costs are all-inclusive – tax and land excluded)

	1996	2001	2006
Newly constructed offices (m² LFA)	0.7 min	1.9 min	0.6 min
Gross rent/m² LFA	145	150	135–160
Net rent/m² GFA	100	105	90–110
Net initial yield (%)	6.5–7.0	6.0–6.5	5.5–6.0
Capital value/m² GFA	1450–1550	1600–1750	1650–2000
Building costs/m² GFA (in 2004 prices)	1250–1300	1350–1450	1400–1500
Residual land value/m² GFA	200–250	250–300	250–500

in order to acquire a lease contract in times of oversupply. This is illustrated in Table 12.1 as a variance of gross rent in 2006 from €135 to the published level of €160 (2004 prices). A NIY in the market of 5.5 per cent in 2006 was an over-reaction. More appropriate in relation with actual real interest and growing risks was an upper yield of 6 per cent.

The situation for a medium-sized office development can, therefore, be described in the following terms:

- Contracted rents are inelastic for diminishing initial yields and changing market activity.
- Capital values of office property are primarily accelerated by diminishing yields.
- Residual land values relatively grow fast, but absolutely decline after 2001. (That created problems for the Land Departments of the Dutch municipalities, who profited from a high non-residential building activity in the late 1990s.)
- Note that unpublished lease-free periods and other incentives are not calculated in property values.

Owner-occupiers of new built offices are probably not better off. They have to compete with other owner-occupiers and investors to acquire their office property. That means, in a growing market, that they have to purchase at the capital values of the completed product. In the shrinking market since 2001, they have to bargain for lower land prices, lower building prices and attractive finance arrangements. This is outside real estate statistics. An important notion is that the demand for office replacement absorbs the newly constructed supply which is available for expansion of the stock.

The general real estate market shows comparable behaviour: inelasticity of rent and real annual payments for changes in building activity and changes in real interest. A higher building activity is accompanied by higher costs of building and land. The real estate industry captures the higher capital values. After

2001, real building prices came under downward pressure, but these have been rising again since 2005. Later in the chapter, this will be analysed for residential development. In the next section, attention is given to the annualized building costs of non-residential real estate.

Annualized building costs

The index real annual payment in the lower part of Figure 12.4 combines the loop over time of the actual real mortgage interest with the 'index real building prices' (construction costs). The annuity is based on a 30-year pay-back period of the original investment. The index for real annual payment (1987 = 100), is a multiplication of the index building prices and the index real annuity. (The appendix to this chapter contains a table showing nominal interest rate, inflation, real interest rate, real 30-year annuity, index of real building prices and the index of real annual payments 1969–1907. Figure 12.4 is based on this table.)

Under the influence of the rising real interest rate and the rising real building prices, the costs (for finance) of buildings rose rapidly in the late 1970s. Therefore, buildings became more expensive. Low investment in buildings in the early 1980s was also related to an interest peak with an interest rate on mortgages of 11 per cent in 1981 (Figure 12.4). Nominal (mortgage) interest and inflation declined since 1982, but real interest climbed to reach nearly 8 per cent in 1987. Real annual payment definitively decreased after 1987. This was primarily influenced by the decreasing real interest rate.

The affordability and feasibility of real estate investment is mostly linked to high nominal mortgage interest, which peaked in 1974, 1982 and 1991. Future inflation and investment calculation on the basis of inflation hedging weakens the influence of nominal mortgage interest on investment initiatives. The line for the real mortgage interest shows the opposite. A real mortgage interest peak is shown in 1987 because inflation was incidentally about zero. Real mortgage interest, actual nominal mortgage interest corrected for actual inflation, abstracts from expected inflation during the operating period of the building. Calculating with a 30-year annuity then provides for a smoothing of the real interest effect on investment.

The decline of non-residential building in the early 1990s was in combination with high nominal and real interest levels and diminishing economic growth. This decline was an overreaction of the market. From 1995 onwards, all was positive for real estate investment. Nominal mortgage interest declined from about 7 per cent in 1995 to about 5 per cent in 2001 and real interest from about 5 per cent to 1 per cent in 2001. Economic growth was high with more than 4 per cent growth in 1998–1999. The shortage of office buildings, in particular, was also high and development for the free market could be started without financial loss due to risk.

Regression analysis led to the conclusion that the index real annual payment is insufficient to explain and to predict investment growth. The correlation and

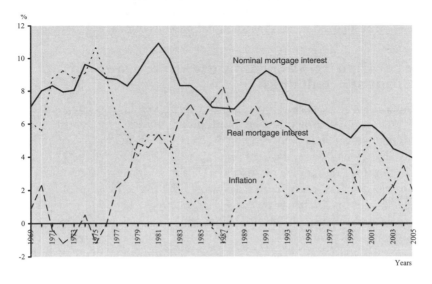

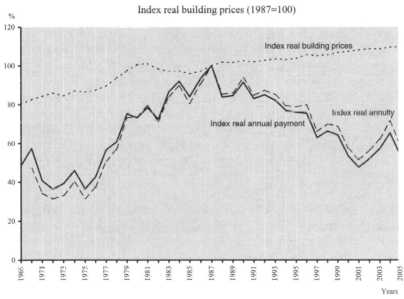

Figure 12.4 Annualized capital and building costs.

Source: Statistics Netherlands (CBS)/elaboration TU Delft – Building Economics.

significance for the period 1980–2005 is low. GDP growth is predominant for explaining past investment and predicting future investment in buildings.

Acceleration from economic growth to growth of investment in buildings

In Figure 12.5, the cyclical loop of investment in buildings is confronted with GDP growth. GDP growth means more production capacity, to some extent more employment and more spending on construction and the use of buildings. When GDP growth is low, the need for expansion of macro-production capacity and of the stock of buildings is also low. Therefore, the investment for expansion of the stock decreases. Zero stock growth means no investment for expansion. This effect is known as the 'acceleration effect'. The lowest points in the economic growth path, for instance 1981, 1993 and 2003, are combined with a decrease in investment.

The regression for $\Delta INV_{buildings}$ = 3.76 × GDP growth (5 years smoothed) − 8.51 per cent, with a R-square of 0.68. GDP growth is significant at 99 per cent.

Note the deviation in 1993, where the regression lags behind the investment activity. This demonstrates an overreaction on diminishing economic growth. Ten years later, around 2003, the smoothed economic growth is on its lowest level since the early 1980s and the investment modelling is close to the actual figures.

Over time, this kind of regression does not provide for a valid forecast of the investment changes because there is a changing relation between the stock of buildings and annual construction and capital formation. A growing stock means that, over time, replacement of stock and other renewal activities within the existing stock become more important. Here the investment pattern deviates from investment in expansion, which is primarily sensitive to the acceleration effect.

The relationship between investment in buildings and GDP is more complicated than an acceleration effect. Figure 12.6 illustrates the broader relationship. Further analysis shows that investment in buildings is linked with the absolute growth of the GDP. That is the investment for expansion of the stock of buildings. The residual part of the investment in buildings is identified as investment for renewal.

In Figure 12.7, the regression of investment in buildings, in relation with the absolute growth of GDP, is plotted as indicator. The best regression is found for the years 1982 until 2005, with a time lag of 3 years.

$INV_{buildings}$ = 0.43 × ΔGDP_{t-3} (5 years smoothed) + €10.100 mln, with a R-square of 0.90 and with 99 per cent significance.

The indicator in Figure 12.7 shows a best fit for the 1990s and an overreaction after 2001. The extrapolation of the trend on the basis of future 2.1 per cent GDP growth is constructed by neglecting the dip in economic growth around 2003.

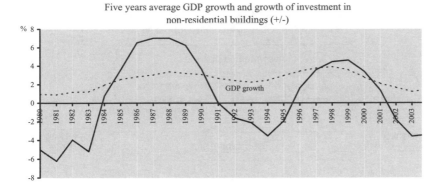

Five years average GDP growth and growth of investment in
non-residential buildings (+/-)

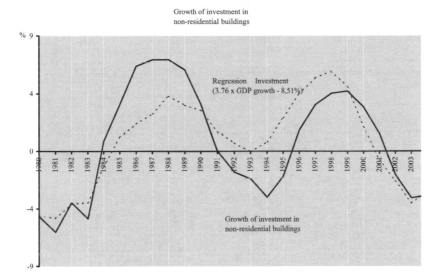

Figure 12.5 GDP–building acceleration.

Source: Statistics Netherlands (CBS)/elaboration TU Delft – Building Economics.

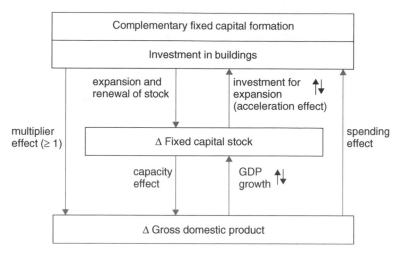

Figure 12.6 Cross effects between GDP and fixed investment.

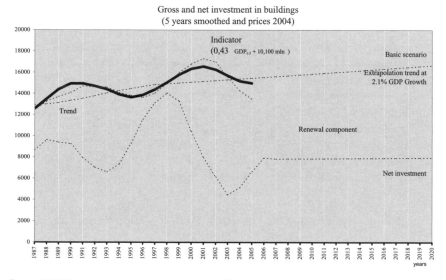

Figure 12.7 Gross and net investment in buildings.

Source: Statistics Netherlands (CBS)/elaboration TU Delft – Building Economics.

Gross investment and net investment

Macroeconomics distinguishes between gross and net capital formation. The difference is the depreciation on the domestic fixed capital stock. Here, we are interested in what kind of investment, with what physical and functional characteristics, is behind the administrative recorded depreciation. The difference between gross and net investment is otherwise labelled as investment for renewal of stock, including replacement.

In the case of non-residential buildings, investment for expansion can be allocated to the absolute growth of GDP. Detailed analysis behind this approach indicates that, in the second half of the 1990s, the most extreme shortage of stock was visible. In Figure 12.7, this is translated by a nearly full allocation of the investment in buildings to expansion (net investment). The link with the absolute growth of GDP is made under the additional assumption that buildings productivity grew 0.9 per cent p.a. since the 1960s.

In Figure 12.7, the upper line is the annual gross investment in non-residential buildings. The lower line is the net investment, which is linked with a smoothed (5-year average) GDP growth. The gap between gross and net is available for replacement and other renewal (renewal part). In the years 1996–1998, nearly all investment is linked with an extreme shortage of buildings at that time. The net investment line is projected until 2020 and is the combined effect of the projected economic growth (2.1 per cent p.a.) and a growth of buildings productivity of 0.9 per cent p.a.

In Figure 12.8, this is turned around and extended to 2020. The lower line now reflects the renewal part. From the starting point the trend is again based on a 21-year average and is completed by extrapolation on the basis of the total trend 25 years before. This renewal line converts to the investment for renewal on a life-cycle basis. Detailed research led to the conclusion that Dutch building investment, analysed in the way described, tends to an investment turnover period of 35 years. This is combined with the additional assumption that investment for extension of stock is fully repeated after 35 years and investment for renewal for 50 per cent. This is an overestimation of the investment for renewal and should be interpreted as an upper limit. The trend of total gross investment in buildings is also based on a 21-year average and for the period 1997–2020 completed as the sum of the renewal trend and the net investment trend (Figure 12.8).

Extrapolation on the basis of regression (Figure 12.8) leads to underestimation because investment for renewal gradually will grow. This leads to a higher loop of the trend in Figure 12.8.

In this chapter, the projected GDP growth is 2.1 per cent in the basic scenario, which is based on the actual trend of absolute GDP growth. The 2.1 per cent figure fits with expectations of Dutch economy in the long run. A constant or slightly diminishing labour force leads to an economic growth which equals macro labour productivity.

An investment turnover of 35 years may not be understood as elimination after 35 years, but as a re-investment after (on average) 35 years, varying from major maintenance to incidentally new construction for replacement. The growing renewal is an echo effect of the huge expansion of stock around 1970. We base our modelling for renewal on a marginal circulation time of 35 years. An additional research question is what will be the 'renewal of the renewal'.

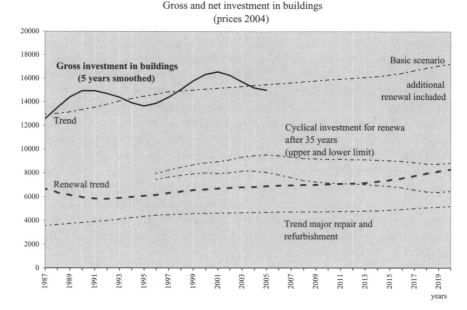

Figure 12.8 Investment in renewal.

Source: Statistics Netherlands (CBS)/Economic Institute Construction Industry (EIB)/ elaboration TU Delft – Building Economics.

At this moment we calculate for a second re-investment of 50 per cent of first re-investment. Further research is still needed in this area.

In practice, the choice is between renewal by replacement and renewal by renovation, refurbishment and major maintenance. The line for investment in existing buildings is plotted in Figure 12.8. This type of investment is an important part of the total investment in buildings. The difference between calculated investment for renewal and investment in existing buildings is invest-ment for replacement. Other investment in new buildings is, in the macro view, investment for expansion of the stock.

Market segmentation – new/old property

Investment in existing buildings became more popular and officially accepted since the 1970s. In general, these investments are important for a sustainable approach to the investment task. Nevertheless, the modernization of the build-ing stock after the 1980s crises was primarily realized by new construction. At present, the huge level of new construction in former years creates a growing vacancy of second-hand buildings and a growing deterioration of older industrial and logistic sites.

The need for modernization and relocation of the stock is evident. When the existing stock runs out of use and becomes vacant, we have to look more systematically for reuse in its original or in a second function. Nevertheless the reuse of the existing stock also meets problems in terms of costs and benefits and in relation to location and negative externalities. The Dutch Office Market since 2001 exemplifies, in an extreme way, the new/old segmentation. .

Figure 12.9 shows a split of total supply in new and secondhand supply. After 2000, a lower absorption of offices, in combination with pipe line effects in new construction, leads to growing supply of new offices. The supply of new offices diminishes after 2002. Three shifts become manifest:

- From expansion to replacement of office stock.
- From vacancy in new buildings to (structural) vacancy in older buildings.
- From supply of new offices to secondhand supply.

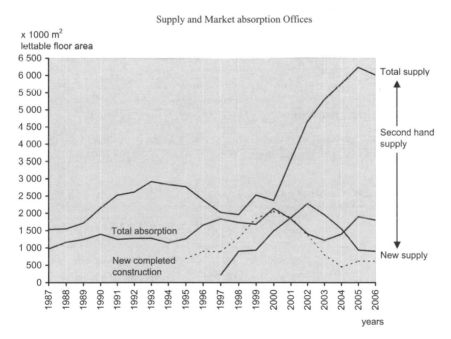

Figure 12.9 Unbalanced office market.

Source: Bak/Vastgoedmarkt/PropertyNL/DTZ/elaboration TU Delft – Building Economics.

The stock adjustment process is, therefore, out of balance. The outcome in terms of office stock is summarized in Table 12.2.

In 2007/2008, there is again a moderate expansion of the stock in use, a growing demand for replacement and at the same moment a vacancy rate of 13–14 per cent of the total office stock. Prices and capital values of vacant buildings are downward inelastic. Investors accept a longer vacancy period. Demolition and transformation to other functions are limited activities.

What are the market interdependencies in the office sector? In the office space market, oversupply leads in general to conditions for a buyer's market. There is an over-supply of old buildings, but, at the same time, a shortage of modern buildings. There are favoured opportunities for betterment, nevertheless users prefer replacement by modern buildings. The office market is out of balance.

In the property/asset market, low real interest means low costs of ownership. In the Netherlands approximately 60 per cent of the office stock is rented and 40 per cent is owner-occupied. Contracted real rents are downward inelastic. Investors prefer high contracted rents in combination with rent-free periods and incentives. Investors seem to neglect long-term risks. They are content with high capital returns (growth of capital values) and moderate income returns out of net rent. The development and construction market after 2001 became a difficult market. Development for the free market is diminishing, because of high risks. Residential development (especially of owner-occupied dwellings) is an alternative development market. The margins between revenues and costs are shrinking and there is latent shortage of land capacity for development and of labour capacity for construction.

Construction market

Within the microeconomic framework for the market equilibrium of demand and supply and of prices and costs, the construction market conditions are

Table 12.2 Lettable office stock (million m²)

	Beginning of 2001	Beginning of 2007	2007 and after
Available stock	40.5	44	Recovery new construction
Stock in use	37.5	38	Further growth dependent on office employment growth
Total vacancy	3	6	Growing gap between available stock and stock in use
Vacancy (new buildings)	1.5	1	Less free development. More pre-contracts and more offices under construction (1 million m²)
Vacancy (old buildings)	1.5	5	Lack of alternative use

illustrated in Figure 12.10. Line ED illustrates the demand function in the period 1998–2001. GB shows the demand function in the period after 2001. ABCD reflects the supply function and is based on the marginal cost of construction. Between A and B, the supply–cost relationship is rather elastic. In section BCD, the costs are rising. This is influenced by lower efficiency, lower productivity and limited capacity of the construction industry when production is expanding. This effect was manifest in the years around 2000 when construction output passed the capacity bottlenecks.

PB (actual market pricing), which is lower than PC (long-term full-cost pricing), illustrates the tendency to marginal cost pricing which is manifest on the Dutch construction market in the years 2002–2005. Since 2006, the upward shift has been repeated. Detailed price and cost data are available for residential buildings activity.

The registered price index of new constructed dwellings diverges from the factor cost index in the period 1995–2003 (see Figure 12.11). For developers this is an 'easy market'. Generally speaking, they are operating in a seller's market. In 2004, the prices developers could realize were diminishing. In the residential market, demand is favoured by low interest rates and, in the Dutch setting, a €1 interest payment on mortgages is 'subsidized' by a €0.42–0.52 tax reduction. This can otherwise be understood as roughly half the interest percentage on mortgages after tax and provides for a high capitalization rate and a high borrowing capacity of the owner-occupiers.

For developers in the non-residential sector, the margin between selling revenues of buildings (after land costs) and construction costs is no longer growing. Selling prices can be considered as gross initial rent divided by gross initial yield. Gross initial rent is under tension from buyers' market conditions. The

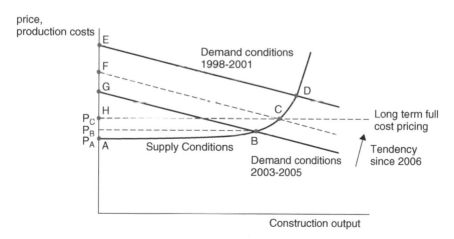

Figure 12.10 Hypothetical demand–supply scheme in relation to long-term full-cost pricing.

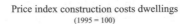

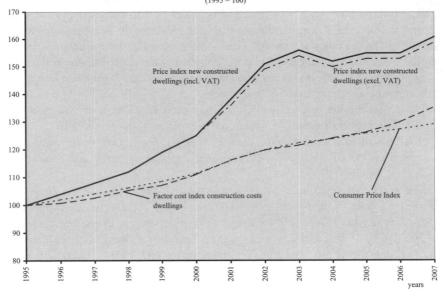

Figure 12.11 Price index of new dwellings.

Source: Statistics Netherlands (CBS)/elaboration TU Delft – Building Economics.

required yield can be adapted as a historical low real interest. Making such calculations is full of risk because even new rental buildings suffer from growing vacancy losses which should be corrected by a higher yield instead of a lower yield, but a higher yield is not marketable. In the office sector, this contributed to a dramatic downswing of new constructed office buildings, from more than 2 billion m² in 2001 to 0.5–1 billion in later years. That is why Dutch developers are nowadays more interested in developing dwellings instead of offices.

Generally, real estate development (residential and non-residential) has to account for smaller margins between selling revenues and development costs. Construction firms try to compensate their loss of development possibilities by more acquisitions on the general construction market (as contractors). Since 2006, production capacity has again been limited by the available labour input. Real factor prices rise and this lowers the profit margin.

Conclusions

For development and construction firms, the market is switching from expansion to renewal of the stock and from past dominance of seller's market conditions to future dominance of buyer's market conditions. In the short run, residential development offers an alternative for lower non-residential activity, especially the decline of new office construction. Project and urban development are threatened by growing risks. The market for the coming decade is more for the renewal of stock and a stabilization of the expansion of the stock of buildings. The demand side of the construction market tends to new combinations of design, build, finance, management and operate. The supply side of the construction market will be handicapped from latent shortage of labour capacity. The European call for a competitive market has to be met by a proactive operating construction sector.

References

Bak, R.L. (2003–2007) *Kantoren in cijfers* (Offices in numbers). CBRE: Amsterdam

Brounen, D. and Eicholtz, P. (2004) *Demographics and the international office markets; consequences for property portfolio*. Rotterdam/Maastricht.

CBS (Statistics Netherlands) (1973–2004) *Nationale Rekeningen* (National Accounts for the Netherlands). SDU: The Hague.

CBS (Statistics Netherlands) (2006) *StatLine databank*, CBS. http://statline.cbs.nl. (accessed December 2007).

CPB (2007) *Macro Economische Verkenningen 2007* (Macro Economic Outlook 2007). SDU: The Hague.

Di Pasquale, D. and Wheaton, W.C. (1996) *Urban Economics and Real Estate Markets*. Prentice-Hall: New Jersey.

EIB, Economic Institute for the Building Industry (annual) *Verwachtingen bouwproductie en werkgelegenheid* (Expectations building production and employment). EIB: Amsterdam.

EIB, Economic Institute for the Building Industry (2006) *Vraag naar kantoren tot 2015* (Demand for offices until 2015). EIB: Amsterdam.

Geltner, D.M., Miller, N.G., Clayton, J. and Eichholtz, P. (2007) *Commercial Real Estate Analysis and Investments*. Mason: Ohio.

Property NL, Quarterly volumes (Several dates).

Vastgoedmarkt (Real estate market): Monthly volumes (Several dates).

Appendix

	Nominal mortgage interest (%)	Inflation (%)	Real mortgage interest (%)	Real 30-year annuity (%)	Index real 30-year annuity (1987 = 100)	Index real building prices (1987 = 100)	Index real annual payment (1987 = 100)
1969	7.05	0.0	7.05	7.07	91.4	80.3	73.4
1970	8.01	5.6	2.28	4.43	57.3	82.7	47.4
1971	8.32	8.8	−0.42	3.15	40.7	84.3	34.3
1972	7.94	9.2	−1.20	2.82	36.4	86.0	31.3
1973	8.04	8.8	−0.67	3.04	39.3	84.5	33.2
1974	9.62	9.1	0.49	3.56	46.0	87.1	40.1
1975	9.33	10.6	−1.18	2.82	36.5	86.4	31.6
1976	8.78	8.9	−0.07	3.30	42.7	87.4	37.3
1977	8.71	6.4	2.18	4.38	56.7	89.6	50.8
1978	8.31	5.4	2.79	4.70	60.8	93.7	56.9
1979	9.12	4.1	4.85	5.82	75.2	97.4	73.3
1980	10.15	5.3	4.57	5.66	73.2	100.7	73.7
1981	10.92	5.3	5.30	6.07	78.5	101.2	79.4
1982	9.96	5.3	4.46	5.60	72.4	98.3	71.2
1983	8.33	1.8	6.40	6.69	86.5	97.2	84.1
1984	8.33	1.1	7.14	7.12	92.0	97.4	89.6
1985	7.77	1.6	6.06	6.50	84.0	95.9	80.6
1986	7.00	−0.3	7.33	7.23	93.4	97.2	90.8
1987	6.95	−1.2	8.21	7.73	100.0	100.0	100.0
1988	6.90	0.8	6.03	6.48	83.8	101.8	85.3
1989	7.57	1.3	6.14	6.55	84.6	101.6	86.0
1990	8.72	1.6	7.05	7.07	91.4	102.5	93.7
1991	9.23	3.1	5.93	6.42	83.1	102.0	84.7
1992	8.84	2.5	6.19	6.57	85.0	102.7	87.3
1993	7.50	1.6	5.81	6.36	82.2	103.5	85.1
1994	7.26	2.1	5.09	5.95	77.0	103.1	79.4
1995	7.12	2.1	4.95	5.87	75.9	103.7	78.8
1996	6.25	1.3	4.89	5.84	75.5	105.7	79.8
1997	5.82	2.6	3.10	4.86	62.9	105.0	66.0
1998	5.56	1.9	3.58	5.12	66.2	105.6	69.9
1999	5.14	1.8	3.30	4.97	64.3	106.7	68.6
2000	5.88	4.1	1.69	4.14	53.5	107.3	57.4
2001	5.88	5.1	0.74	3.68	47.5	108.0	51.4
2002	5.33	3.8	1.45	4.02	51.9	108.6	56.4
2003	4.48	2.2	2.25	4.42	57.1	108.6	62.0
2004	4.20	0.7	3.44	5.04	65.2	109.5	71.4
2005	3.90	2.1	1.78	4.18	54.1	109.7	59.3
2006	4.50	1.9	2.53	4.56	59.0	110.9	65.4
2007 (preliminary)	4.50	1.5	2.96	4.79	61.9	113.0	69.9

Chapter 13

Theories of investment in property

Use of information, knowledge and intelligent technologies

Arturas Kaklauskas and Edmundas Zavadskas

Introduction

At first sight, the subject of investment in property looks anything but complicated. However, such an attitude changes after a more thorough analysis of the issues related to this area. Recent books and publications on investment in property (real estate investment) are abundant, and certain topics appear in more than one text. As an example, we briefly list several typical topics that are related to investment in property and repeated in many texts, e.g. investment goals, comfort zone, real estate values and trends, location, land use regulations, investment options (apartment complexes, hotels, office buildings, shopping centres or industrial buildings etc.), microlevel investment analysis, macrolevel real estate investment issues (portfolio theory and institutional landscape etc.), measuring investment performance, management of investment processes, real estate development, risk appraisal techniques, valuation models, real estate transaction, buying techniques, negotiation, forms of ownership available to investors (partnership, corporate ownership, syndication, land trusts, limited partnerships), financing techniques (second mortgage, adjustable-rate mortgage, fixed-rate mortgage, reverse mortgage, balloon mortgage, assumable mortgage, leverage, creative financing techniques), financing strategies (how to be at the right time and at the right place etc.), tax benefits and foreclosure, etc. These issues are dealt with in this chapter.

Investment in a building or investment in a built and human environment?

A built environment is developed in order to satisfy residents' requirements. Human needs can be physiological or social and are related to security, respect and self-expression. People want their built environment to be aesthetically attractive and to be in an accessible place with a well-developed infrastructure, convenient communication access and good roads, and the dwellings should also be comparatively cheap, comfortable, with low maintenance costs and have sound and thermal insulation of walls. People are also interested in ecologically clean and almost noiseless environments, with sufficient options for relaxation, shopping, fast access to work or other destinations and good relationships with neighbours.

It must be admitted that the most serious problems of built environments, e.g. unemployment, vandalism, lack of education, robberies, are not always related to the direct physical structure of housing. Increasing investment into the development of social and recreational centres, such as athletic clubs, physical fitness centres, and family entertainment centres, the infrastructure, a good neighbourhood and better education of young people, can solve such problems.

Investment, purchase and sale of a property, and its registration have related legal issues. The legal system of a country aims to reflect its existing social, economic, political and technical state and the requirements of the market economy.

From the social perspective, the built environment can affect the whole of society or specific groups of people and individuals. For example, poor dwellings are non-aesthetic, uncomfortable and can be sources of various diseases or pose acute social problems such as a dirty or crime-ridden environment. These factors affect neighbourhoods from various perspectives. For example, in transitional countries, some low-income households (retirees, large families or the unemployed) cannot afford to pay for utility services, such as heating and hot water, without state support. By failing to solve such problems at the national level, a government may lose considerable constituency in any ensuing election. Thus, the problem is not only social but it is also political. Similar problems occur when governments attempt to create better conditions for long-term mortgage loans and must therefore intervene in financial markets.

The built environment is not constructed in an empty space. During the built environment life cycle – brief, designing, construction, maintenance, facility management, renovation, demolition and utilization – buildings are affected by various micro, meso and macro level factors.

It is estimated that about 20 per cent of the US population suffers from asthma, emphysema, bronchitis, diabetes or cardiovascular diseases and are thus especially susceptible to external air pollution (American Lung Association, 2005). Outdoor air quality plays an important role in maintaining good human health. Air pollution causes large increases in medical expenses and morbidity and is estimated to cause about 800,000 annual premature deaths globally (Cohen et al., 2005). Much research, digital maps and standards on the health effects (respiratory and cardiovascular effects, cancer, infection etc.) of outdoor air pollution, a premise's microclimate and property valuation, have been published in the last decade. These and other problems, are related to a built environment's air pollution, a premise's microclimate, health effects, and real estate market value.

These examples allow one to come to the conclusion that various stakeholders usually prefer the concept of a 'built and human environment' to the concept of a 'built environment'. This proposition is especially true, when not only the built environment, but also the surrounding micro, meso or macro environment, is considered as a research object.

Currently, a built environment is characterized by the intensive creation and use of information, knowledge and automation applications (software, knowledge, expert and decision support systems, and neural networks). It is commonly agreed that use of these applications will significantly speed up built environment processes, improve the quality of the built environment and the value of decisions made and decrease the overall cost of a built environment's life cycle.

Comfort zone

A comfort zone denotes that limited set of behaviours that a person will engage in without becoming anxious. A comfort zone is a type of mental conditioning that causes a person to create and operate mental boundaries that are not always real. Such boundaries create an unfounded sense of insecurity. For example, inertia is when a person who has established a comfort zone in a particular axis of his/her life, tends to stay within that zone without stepping outside of it. To step outside a person's comfort zone, he/she must experiment with new and different behaviours, and then experience the new and different responses that then occur within his/her environment. The boundaries of a comfort zone can result in an internally rigid state of mind. A comfort zone may alternatively be described by such terms as rigidity, limits or boundaries, or a habit, or even as stigmatized behaviour (Bardwick, 1995).

For example, the following is information that a real estate investor should strive to learn regarding the investment's comfort zone (Quadreal, 2007): geographic layout; street names; subdivision names; zoning rules and regulation; local ordinances that affect real estate; price ranges by subdivision or streets; rental market data and rents charged; future road plans; future utility plans; future developments in planning stages; local employment statistics; employment trends; major impacts, that will affect employment trends; the 'how' and 'who' of the local government; 'what', 'how' and 'who' of the local building department; school districts and how to enter other schools; bus and other local transport routes; 'what', 'who' and 'how' of public records; names of prominent business leaders in the community; and sources for local financing. The investor should keep in mind that many of these factors change from time to time and thus a constant review of the current circumstances is required.

Consideration of investment techniques is also required, i.e. the option agreement, the lease option agreement, wrap-around mortgages and secondary seller-held financing etc. An investor can find that investment will be short lived if he/she needs to rely on conventional methods of purchasing and financing the properties that he/she find in his/her comfort zone. These techniques will become the tools for building an investment portfolio (Quadreal, 2007).

Real estate investment and portfolio theory

Real estate investment involves the purchase of real estate for profit. Profits are accumulated slowly by renting out properties in a cashflow method, or properties are generally improved and resold for a capital gain. The biggest factor in the marketability of an investment is supply and demand. Leverage, or the ability to borrow based on the value of the property, is probably one of the greatest advantages. It is much easier to finance real estate than any other product. While investment in most assets requires the purchaser to have the full purchase price available for the asset, in real estate investment, one only needs to have a fraction of the purchase price available as a down payment. Therefore, real estate, although incredibly expensive, is still easier to buy than, say, a piece of industrial equipment of the same value. Real estate is a non-liquid investment that needs maintenance and payment of taxes. A balanced investment portfolio has some liquid assets that can be quickly converted to cash to sustain the real estate when its returns are not sufficient to cover its recurring costs (RLI, 2008).

Portfolio theory, originally developed by Harry Markovitz in the early 1950s, sometimes referred to as the modern portfolio theory, provides a mathematical framework in which investors can minimize risk and maximize returns. The central plank of the theory is that diversifying holdings can reduce risk, and that returns are a function of expected risk (Portfolio theory, 2007). The key result of the portfolio theory is that the volatility of a portfolio is less than the weighted average of the volatilities of the securities it contains (Portfolio theory, 2006). The volatility is the standard deviation of the expected return on a security. The volatility therefore changes with the period of time over which it is measured (Volatility, 2007). The expected return on most investments is uncertain. However, it is possible to describe future returns statistically as a probability distribution (Expected return, 2007).

An efficient portfolio is one that lies on the efficient frontier (Efficient portfolio, 2007). The efficient frontier describes the relationship between the return that can be expected from a portfolio and the risk of the portfolio. This can be drawn as a curve on a graph of risk against the expected return of a portfolio. The efficient frontier gives the best return that can be expected for a given level of risk or the lowest level of risk needed to achieve a given expected rate of return (Efficient frontier, 2007). An efficient portfolio provides the lowest level of risk possible for a given level of expected return. An efficient portfolio also provides the best returns achievable for a given level of risk. If a portfolio is efficient it is not possible to construct a portfolio with a higher expected return and the same or a lower level of volatility with the securities available in the market, which excludes risk-free assets (Efficient portfolio, 2007).

Investors can reduce risk and improve the level of risk relative to the return by diversifying their portfolios. The key to diversification is to choose investments whose prices are not strongly correlated. Investing in different sectors, geographical regions and classes of security improves diversification: the values of shares, bonds and pieces of real estate will be more correlated with each other than with investments of completely different types (Diversification, 2007).

Life cycle portfolio models are designed to identify optimal savings and portfolio policies over the lifetime of investors. The standard portfolio theory, introduced by Markowitz (1952), is static in nature, since it explores investment decisions for only one period. A more realistic setting must account for the multi-period dimension of the portfolio choice problem. Only under very specific circumstances is the optimal portfolio structure time invariant. In this special case, a one-period optimization suffices to characterize the optimal portfolio choice also in a multi-period environment. Under more general conditions, however, investors will restructure their portfolios in reaction to changes of income, the accumulated wealth and the investment opportunity set. This opportunity to adjust the portfolio composition affects the initial investment choice (Wallmeier and Zainhofer, 2006). Sufficient software and intelligent systems have been developed for real estate investment and several are briefly described here.

Real Estate Offer Generator (2007) is real estate software that calculates the price offered for a rental property. The generator can help users to buy properties that make for a positive cash flow. The software has an easy-to-use interface in order to help investors calculate the net operating income, create various offers and the initial offer letter, as well as calculate the cash flow and projected cash flow for each of their options. The software allows professional and amateur investors to gain a strong advantage in real estate investment and move on a solid mathematical basis.

The classical Markowitz Sharpe optimization model for investment portfolios is applied in practice through the Real Estate Offer Generator. The software allows a customer to import market data, define groups of assets, specify legal and market constraints and then find the optimum portfolio composition. Real Estate Notebook (2007) can assist real estate investors to analyse and organize investment properties. The software performs calculations that are crucial to property analysis including mortgage amortization, total expenses, return on investment, net operating income, depreciation and many other aspects. The Real Estate Notebook can store all analysed properties for later viewing or reporting based on criteria that the individual specifies. The software includes a unique portfolio of reporting features that show the performance of an individual real estate portfolio as a whole and as a charting feature for quick visual head-to-head analysis of properties.

Software for real estate investment (2007) produces projections and presentations of up to 20 years for office buildings, industrial buildings, shopping centres, apartments and mixed-use properties. A customer can forecast the commercial revenue stream in detail, as well as operate expenses, pass-throughs, financing, cash flows, tax liability, resale, and rates of return and partnership allocations. Software for real estate investment (2007) has also released two optional add-on products that allow one to compare multiple investment property scenarios and to perform portfolio analysis.

Real Estate Tracker (2007) was created to help customers make intelligent and accurate choices for residential investment properties and integrate the real estate portfolio with a tool to track customer income and expenses in an easy to use budget tracker. The Real Estate Tracker is an online property investment tool designed by investors for investors to empower a customer with information needed to leverage a return on the investment and accurately track the real estate's cash flow over time. The Real Estate Tracker (2007) can further help to identify the best properties to buy, identify the right time is to sell or when to do a tax-deferred exchange, when to raise rent, and alert the customer as to when he/she should pull equity to purchase new investments.

Mortgage and e-mortgage

A mortgage is an instrument for lending money on real estate. The property is pledged as security for the loan, and the lender has the right to take over the property if the borrower defaults on the terms of the loan. Mortgage derives from two French words meaning dead and pledge, because when the loan has been repaid, the mortgage is considered void or dead (Grass, 2007).

There are many types of mortgage loan. The two basic types of amortized loans are the fixed-rate mortgage (FRM) and adjustable-rate mortgage (ARM). In an FRM, the interest rate, and hence monthly payment, remains fixed for the life (or the term) of the loan. In an ARM, the interest rate is fixed for a period of time, after which it will periodically (annually or monthly) adjust up or down to some market index. Adjustable rates transfer part of the interest rate risk from the lender to the borrower, and thus are widely used where unpredictable interest rates make fixed-rate loans difficult to obtain. In most scenarios, savings from an ARM outweigh its risks, making them an attractive option for people who are planning to keep a mortgage for 10 years or less. Additionally, lenders rely on credit reports and credit scores derived from them. The higher the score, the more creditworthy the borrower is assumed to be. Favourable interest rates are offered to buyers with high scores. Lower scores indicate higher risk to the lender, and lenders require higher interest rates in such scenarios to compensate for the increased risk (Patrick, 2007).

There are essentially two types of legal mortgage: a mortgage by demise and a mortgage by legal charge. In a mortgage by demise, the creditor becomes the owner of the mortgaged property until the loan is repaid in full (known as 'redemption'). This kind of mortgage takes the form of a conveyance of the property to the creditor, with a condition that the property will be returned on redemption. This is an older form of legal mortgage and is less common than a mortgage by legal charge. In a mortgage by legal charge, the debtor remains the legal owner of the property, but the creditor gains sufficient rights over it to enable them to enforce their security, such as a right to take possession of the property or sell it. To protect the lender, a mortgage by legal charge is usually recorded in a public register (Mortgage, 2007).

Leece (1997) reviewed developments in the design and innovation of mortgage instruments in the UK, from the early to mid-1990s. Rasmussen et al. (1997) present a more expansive view of reverse mortgages as a financial tool for tapping housing equity for various purposes and at various stages in the life cycle. Van Dyk (1995) examines the mechanisms used since the 1970s to finance social housing in Canada. He demonstrates that direct government assistance has proved to be the most cost-effective mechanism. Experimentation with alternative mortgage instruments, such as graduated-payment mortgage and index-linked mortgage, has also been central to attempts to minimize subsidy and financing costs. Dhillon et al. (1990) evaluate the choice between 15-year and 30-year fixed rate contracts in the US and estimate a simple profit to represent this choice. Lam et al. (1998) developed a model for financial decision making, which provides a method for solving borrowing decision problems. Leece (2000) estimates reduced form credit demand equations that reflect the interactions between the choice of mortgage instrument, the lessening of mortgage rationing and liquidity constraints and the demand for housing debt. Most of these studies have concentrated on single objective decision making.

Housing finance systems differ greatly from country to country. As Renaud (1999) stated, there are profound differences among the 180 developed and developing countries that are now members of the World Bank. Advanced housing finance systems can be found in OECD countries. Renaud (1999) shows that these systems grew out of two main traditions: Anglo-Saxon systems where the building societies of the UK and the savings and loans from the US are mutual forms of housing finance. There is also the mortgage bank tradition of continental Europe where term funding was mobilized through bond markets.

There has been an increase in literature on the choice of housing investment instruments. This mainly concerns the econometric estimation of the demand for fixed rate mortgages compared with adjustable rate mortgages. There is also a large amount of empirical work on the mortgage choices between the conventional annuity mortgage and payment via savings in a diversified portfolio of assets.

When a homeowner defaults by failing to make payments on his/her mortgage, the bank or financial institution that holds the mortgage note can foreclose on the property. Foreclosure gives the legal ownership of a property to the bank, so as to allow the bank to recoup its investment. Foreclosure proceedings vary but usually involve court appearances to ensure that the foreclosure is warranted. Pre-foreclosure sales can allow a defaulting borrower to sell the mortgaged property to satisfy the loan and avoid foreclosure. The proceeds of the sale are used to pay the mortgage debt, with any excess going to the mortgagor or property owner (Schwartz, 2007).

In general terms, the main participants in a mortgage are creditor, debtor and other participants such as the mortgage broker and financial adviser. The creditor has legal rights to the debt secured by the mortgage and often makes a loan to the debtor of the purchase money for the property. Typically, creditors are banks, insurers or other financial institutions that make loans available for the purpose of real estate purchase. The debtor or debtors must meet the requirements of the mortgage conditions (and often the loan conditions) imposed by the creditor in order to avoid the creditor enacting provisions of the mortgage to recover the debt. Typically, the debtors will be the individual home owners, landlords or businesses that are purchasing their property by way of a loan. Due to the complicated legal exchange, or conveyance, of the property, one or both of the main participants are likely to require legal representation (Patrick, 2007).

Clearly, the Internet is poised to have a significant impact on real estate capital markets, serving primarily as a new platform for the delivery of data and services. Through on-line mortgage firms, real estate finance seekers can easily find information about mortgages, rates, fees, duration and upcoming offerings. Closing a deal is still a traditional transaction. In order to solve real estate finance issues more efficiently, virtual loan and financing markets are created. Developers, brokers, investors and lenders are involved in these activities. Such virtual loan and financing markets contain real estate software and intelligent systems that facilitate their activities.

When considering applying for a loan with a mortgage company, an investor should determine the following: interest rate, time required for approval and closing the deal; ease of loan service; and familiarity with and perceived professional competencies of the loan staff. After a lender's selection, a mortgage package is required for submission with the mortgage application, so that the lender can approve it. The mortgage package contains many items: e.g. a mortgage application listing the amount of loan requested; personal financial data of the borrower; the borrower's job history; real estate to be purchased; and the agreed sales price. Further, it contains a verification of employment and salary; credit checks; real estate appraisal; and verification of bank deposits and/or loan amounts etc. There are many on-line mortgage brokers. However, although many consumers are researching loans on-line, few are closing loans through the Web.

One of the most important goals of a potential home buyer, is to find the best possible option for credit. It can be claimed that discovery of the best possible loan equals in importance the discovery of the best dwelling for lower-income households, who not only become home owners but also assume a serious financial commitment. Types of loans are abundant and a search for, and assessment of, all of them is a rather complicated process for the consumer. The so-called sector of intermediaries, deals with these issues and their aims are to put in touch those who demand and those who offer/supply loans. This search for a rational loan life cycle can be very expensive. The process of searching for loans and filling in application forms includes human and bureaucratic expenditures and increases the cost several times as compared to using an e-loan.

Some companies are already offering online services which allow clients to search for and get loans from numerous alternative creditors. By sending queries, customers can get comparisons of interest rates and charges, and then the most appropriate option can be selected by a mouse-click. The loan is perceived as a commodity and its selection is primarily based on determination of buyer's needs i.e. those who seek a rational loan life cycle and in finding the seller, who can supply the loan or the person, who best meets the needs. This process is Internet-friendly and is more efficient and effective than services of traditional agents. In order to be able to compare conditions offered by various credit suppliers more easily, it must be granted that the information provided by the suppliers is as precise as possible and unified nationwide. Otherwise, if one of the loan suppliers provides incorrect or incomplete data about the offered loan package, a consumer can be misled and the selected loan will not be the most rational.

Explicit and tacit knowledge in a real estate investment

By finding, capturing and sharing explicit and tacit knowledge, investors can significantly improve results. One of the main roles of explicit and tacit knowledge management in real estate investment is sharing best practice. Throughout the world, there are many examples of the adoption of the best practice (investing process, appraisal services, brokerage, consulting, insurance, matching/listing services, mortgages, project development, real estate finance, and real estate's transaction process) by the major players in real estate investment.

Explicit knowledge comprises documents (e.g. investment appraisal, feasibility study of an investment project, balance sheets, buy–sell agreements, insurances, market analysis, contracts and declarations etc.) as well as data that is stored within the computer's memory. This knowledge must be easily accessible, so that an investor can receive all the necessary information without disturbances or inconvenience. Explicit knowledge includes information that is widely used in information technologies.

Knowledge is the integrated sum of physically intangible resources, the bigger part of which is tacit (e.g. skills, competences, experiences, organizational culture, informal organizational communication networks and intellectual capital of an organization). It is frequently believed that the utmost knowledge resource leaves the organization at the end of each working day in the heads of its employees. Capturing the tacit knowledge of individuals in a way that can be leveraged by companies is perhaps one of the biggest challenges in real estate investment. The main investor knowledge is tacit. The creation and distribution of tacit knowledge therefore requires creativity and competence. Tacit knowledge is a mixture of informal and non-registered procedures, practice, skills, deliberations, subjective insight, intuition and judgment that an investor acquires by virtue of his/her experience and expertise. This knowledge is also vitally important because it defines the abilities and experiences of the investor. Tacit knowledge represents an important intellectual resource that cannot easily be duplicated by competitors. Tacit knowledge must be converted into explicit knowledge so that it can be recorded. Recorded knowledge is static and can soon become outdated. Innovative organizations establish an environment where knowledge is continuously created, captured and disseminated.

The transfer of tacit knowledge is unverifiable and requires face-to-face contact, and the creation of spatial nearness is significant. Experts can share information about a current investment issue, problem or topic through meetings, workshops, seminars, video conferencing, e-mail and intranet based on discussion groups, extranets, telephone, working on joint projects, coffee conversations, canteen discussions and brainstorming sessions.

Different knowledge-capture techniques (including interview, on-site observation, brainstorming, protocol analysis, consensus decision making, nominal-group technique, Delphi method, concept mapping) can be used to capture tacit knowledge and writing down tacit knowledge in the form of investment appraisal, feasibility study of an investment project, buy–sell agreements, market analysis, contracts and other methodology. Once knowledge is captured or codified it is no longer tacit.

Best practice in a real estate investment

Much more attention has to be paid to knowledge creation and its distribution in the form of the knowledge and databases of best practice, and this has recently begun in the most progressive activities of real estate investment. Throughout the world, there are many examples of the adoption of best practice by the major players in real estate investment. Some of their works are presented in the following list: comfort zone, land use regulation, choosing investments, investment analysis at microlevel, macrolevel real estate investment issues, measuring investment performance, management of investment process, real estate development, risk, appraisal techniques, valuation models, real estate transaction, buying techniques, negotiation, forms of ownership, financing techniques, financing strategies etc.

Search, storage, management and improvement of best practice, as well as best practice knowledge and databases created on their basis, are some of the newest priorities of real estate investment in most advanced countries. Comparative analyses of best practice are becoming more popular in real estate investment. Comparative analyses are based on the analysis of the best examples of services available to clients. On the basis of comparative analysis, certain recommendations are then formed, indicating how to provide services of a higher quality and how to better serve the needs of clients. Comparative analyses provide the possibility to quickly and efficiently understand and apply methods which will help to achieve the quality of client service at a world-class level.

The best practice in real estate investment is obtained in different ways, e.g. applied research, wisdom and experience stored by practices, experiences of clients and other stakeholders and opinions of experts etc. Databases and knowledge bases of best practice are knowledge-obtaining tools, which allow one to save a lot of time by providing information on the best real estate investment practice in different forms, e.g. studies, reports, agreements, market analysis, contracts, declarations, e-mail messages, slide presentations, text, video and audio material.

Stakeholders most often try to achieve different economic, comfort, technical, technological, social, political and other aims. Different means can be used to achieve these aims. Some aims are not so easy to achieve and others might require more expenditure. Best practice allows one not to limit oneself only to the implementation of economic aims, rather it creates conditions to reach a higher level and realize aims from the perspective of which practice was named as the best one. The main problem of many best practices is the way they are presented, i.e. they are suggested, and do not always take into account a specific or certain situation.

Comparative analysis systems of best practice help investors to determine directions of priority for increasing an activity's efficiency and ways of determining the achieved progress and measuring investment performance, which allow one to compare the performed investments with existing investments, as well as to determine the spheres that are lagging behind and suggest theories and practices of investment in property that will eliminate these gaps. Modern investors know how to use the possibilities of a comparative analysis and therefore decrease their expenditure and increase competitiveness.

Information gathering and comparison intelligent agents

One of the major problems in Web-based information systems is to find what you want. The number of alternative real estate investment products and services on the Internet is in the thousands. How can customers find rational investment products and services on the Internet? Once an investment product or the service information is found, the customer usually wants to compare alternatives. There are specific classes of intelligent agents that search for hypertext files by agents, search alternatives on databases, complete alternative searches and make tabular comparisons, compare alternative products and services from multiple malls, and search multiple criteria decision making.

It can be expensive for brokers and users to find each other. On the Internet, for example, thousands of products are exchanged among millions of people. Brokers can maintain databases of user preferences and suppliers' (i.e. providers') advertisements, and reduce search costs by selectively routing information from suppliers to users. Information-gathering agents, called *worms* and *spiders*, are used to gather information about the necessary contents of the Internet for use in search engines and consume quite a lot of bandwidth in their activities. Information-gathering agents can reduce the waste of band-width. This reduction can be achieved by:

- Using results and experiences of earlier performed tasks to make future executions of the same task more efficient, or even unnecessary. Serious attempts are made where agents share gained experience and useful information with others.
- Using the 'intelligence' of agents to perform tasks outside peak hours, and spreading the load on the Internet more evenly. Furthermore, agents are better at pinpointing at which time of the day there is too much activity on the Internet, especially since this varies between the days of the week as well (Hermans, 2000).

The authors have developed Cooperative Integrated Web-Based Negotiation and Decision Support System for Real Estate (Kaklauskas et al., 2005). A proposed Web-based Intelligent DSS for Real Estate can create value in the following important ways: searching for real estate alternatives; finding out alternatives and making an initial negotiation table; multiple criteria analysis of alternatives; negotiations based on real calculations; determination of the most rational real estate purchase variant.

E-brokerage and e-transactions

Under the traditional system, a real estate agent offers a package of services: showing real estate; advising sellers on how to make the real estate more marketable; assessing current market conditions; providing information about real estate values and neighbourhoods; matching buyers and sellers; negotiating the sale price; signing contracts; arranging for inspections; assisting with closings and so on.

The Internet and intelligent technologies can disaggregate the above services in the following ways: the Internet searches for real estate, finds alternatives and prepares comparative tables, databases that provide information about real estate, their values and neighbourhoods, matches buyers and sellers, negotiates the sale price, assists with real estate selection and lender selection, provides smart software and personalized websites that manage complicated transactions.

Brokers usually work for a commission, acting as intermediaries between buyers and sellers. Brokers are involved in matching, negotiating and contracting. In general, sellers and renters set preliminary prices and these are then negotiated. However, direct negotiations are sometimes undesirable or unfeasible.

Many of the new investments in property portals make economic sense in that they make life better, cheaper or faster for somebody. The greatest real estate opportunity for big profits appears to be in brokerage (both leasing and sales). Brokers, whether human or electronic, can address the following important limitations of privately negotiated transactions:

1 Real estate search costs. The residential brokerage system already has databases in place with shared listings, making transitions to a Web-based system for the sharing of information equitably. Brokers can maintain multiple listing services and reduce search costs by selectively routing information from sellers or renters to consumers and by matching customers/clients with residential buildings. Brokers with access to a customer's preference data can predict demands. Some brokers already offer such services.
2 Lack of privacy. Either the buyer or seller may wish to remain anonymous or at least to protect some information that is relevant to a trade. Brokers can relay messages and make pricing and allocation decisions without revealing the identity of one or both parties.
3 Incomplete information. The buyer may need more information than the seller is able or willing to provide, such as information about a building's quality and the market value. A broker can gather building information from sources other than the building's seller, e.g. independent evaluators.
4 Risk. The broker may accept responsibility for the behaviour of parties in transactions that it arranges and act as an inspector on his/her own.

Ham and Atkinson (2003) focus on five key aspects of the home buying and selling process and discuss barriers to transformation and changes in law and regulations for each:

1 Improving computerized access and accuracy of credit reports by standardizing reporting data to allow for one-stop correction at all credit bureaus and requiring more accountability for accurate reporting of credit history.
2 Facilitating computerized shopping for mortgage interest rates by standardizing forms and eliminating protectionist rules that favour in-state bricks-and-mortar lenders.
3 Unbundling the functions of real estate agents by encouraging competition for brokerage and listing services and disclosing alternatives to buyers and sellers.
4 Streamlining the recording process to cut costs and reduce risks associated with incomplete or inaccurate land records by establishing electronic recording systems.
5 Reducing the costs and paperwork associated with the settlement process by encouraging digital signatures and online settlements.

Many buyers and sellers hire professional agents as the first step in making a sales deal in the real estate market. These agents perform numerous functions: they advise the seller on making the object for sale more attractive for the market; they help to prepare and collect various documents; they represent the client's interests in the negotiations on the price; and they guide him/her through a number of mandatory phases of a real estate deal until moving into the new home. Although these services are really useful, most buyers and sellers claim that the primary reason to hire an agent is to find a suitable dwelling or a buyer/seller. The services of a real estate agent are charged as a commission fee, which is paid by the buyer or the seller and usually makes about 6 per cent of the deal's value. Usually, the seller pays all 6 per cent to its agent, who, in turn, offers part of the amount to another agent who found the buyer (if the buyer is not represented by any other agent, the seller's agent retains all 6 per cent). Information, knowledge and intelligent technologies can reduce these expenditures considerably. Broader application of IT could make prerequisites for a buyer to select only the desired services of an agent, and to leave the remaining services to intelligent technologies.

Online search for home or mortgage also saves the consumer's time; a consumer who makes a search using other than Web-based means undoubtedly wastes more time. Those real estate buyers who search on the Internet can view considerably more potential objects than consumers who use the services of a regular agent. Increased use of IT should also influence the standard commission fee (6 per cent), i.e. the fee which is more related to culture and tradition than based on market logics. Agents provide valuable services, and many buyers and sellers will always prefer the services of an agent, who offers a full service portfolio. However, it must be clients and not agents who should decide what services to buy.

The website of a notary can specify all documents that are needed to complete a deal and which would be available for a thorough analysis of all stakeholders' deals. Each document can be signed by a digital signature and sent via electronic means, saving time and money that is needed to organize meetings. Implementation of this process often requires changing laws and revocation of the mandatory participation of lawyers in the process of deal finalization. Although the client has a right to select his/her own lawyer, this action usually increases the client's expenditures. A transparent and unified e-signature system is required for this purpose. Strict identification of users who use e-signatures should be guaranteed.

The authors have developed Real Estate's Market Value and a Pollution and Health Effects Analysis Decision Support System (Zavadskas et al., 2007). The developed system can create significant value for e-brokerage and e-transactions.

Project development

Many new laws and practices, such as environmental impact reviews, historic preservation requirements, growth controls and impact fees etc., have served to slow the development process and add to the costs of real estate development. Developers find themselves increasingly involved in public relations campaigns and public policy initiatives, working with local residents, business and civic groups, community leaders and government officials to have projects approved by agreeing to pay a greater share for public facilities and amenities. Developers are also busy finding new ways to address neighbourhood concerns and mitigate the perceived negative effects of proposed development. All of these might be decided more easily by using project development websites.

The full-service needs of large projects are now being met by a new generation of websites that integrate virtual community creation, online collaboration and support services to developing an environment in which the whole process from the design stage to the facility management process is running smoothly. These websites bring together investors, designers, economists, building material manufacturers, suppliers, contractors and mortgage brokers involved in project development. Some developers, construction firms and contractors have their own specific project-linked intranets.

In order to increase a project development's efficiency, various software, expert and decision support systems are used. One such computer software system is Commercial/Industrial Development Software. This software performs a complete project cost analysis for any new commercial income property. It also provides the developer with an excellent budget 'pro forma' for presentation to a lender, partner or client. The report includes a project summary and overview, financing and leasing information and a pro forma operating statement and resale projection. The report summarizes land, development, architectural, financing, construction and lease-up costs. Developers, contractors, lenders and others who will be involved in the construction or refurbishing of a commercial, industrial or multi-unit residential income property use this software. Interested parties often use this programme to analyse the development phase and then also use Real Estate investment analysis software to project the performance of the property over time. This software allows one to produce a comprehensive 10-year projection for any type of residential or commercial income property and to construct anything from a simple and straightforward analysis to a highly sophisticated investment scenario. This software serves all who deal with commercial or residential income properties: individual and institutional investors, brokers, appraisers, lenders, attorneys, accountants, portfolio managers, financial planners, as well as architects and developers, etc.

Real estate's transaction process and investment multiple listing service

Steps in a real estate transaction process can be represented by the example of residential transactions. The real estate transaction process can be divided into five stages: listing, searching, evaluation, negotiation and closing transactions. Transaction costs will be reduced directly through a reduction in underwriting costs as appraisals, environmental reviews, title insurance, and other vendors are efficiently contracted and managed through the Internet. A faster and a higher-quality information flow between brokers, owners, lawyers, vendors, lenders and other participants in the transaction process will reduce costs.

A seller may place information about a property intended for sale in various real estate-for-sale databases called a multiple listing service (MLS). Real estate-for-sale databases are operated by the local real estate's broker-board and, on the basis of such databases clients can very quickly find a house they want. MLSs are primarily financed by the sellers, either from the commissions they pay, when they list a house with a real estate broker, or directly to the maintainer of the site. The service lists of real estate for sale and data on sales are made by brokers. Statistical data regarding listings, sales and data about the market and information on the trends are also often provided. MLS data is essential to the professional real estate agent and the appraiser who wants to offer clients a wide variety of available properties and current market data. MLS in most areas represents the vast majority of properties offered for sale. The MLS does an efficient job of quickly selecting specific types of sales in a specific area from the hundreds and thousands of recorded sales. As a rule, sellers try to highlight the positive aspects of a house and suppress drawbacks and defects.

After finding all the possible alternatives, real estate should be assessed. The property needs to be assessed because each buyer has a different understanding of the quality of the real estate. Buyers also pursue their own specific goals. For instance, a buyer may want to have a relatively cheap and comfortable house with low maintenance costs plus good thermal and sound insulation of the walls and a good external aesthetic appearance to the house. Furthermore, he/she desires to have an ecologically clean and quiet environment with good leisure and shopping facilities, good neighbours and excellent transport connections to drive to work or elsewhere. The list of goals pursued by buyers can be extended further. In this area, websites such as Virtual Home Tours (www.hometours.com) offer additional information about houses in the form of a virtual walk-through. Such virtual promenades save both brokers' and potential buyers' time and help the buyer to make a decision on whether or not to take an actual look at the house.

Real estate e-negotiation involves process, behaviour and substance. The process points to how the stakeholders negotiate – context, tactics, stages. Behaviour refers to the relationships among stakeholders, the communication between them and the styles they apply. The substance points to what the stakeholders negotiate over – agenda, interests, options, agreement.

Legally a real estate's ownership is transferred by giving the real estate's deed to the buyer and closing is usually handled by a third party, e.g. a lawyer or the title's company that both parties trust; although, who that is differs from jurisdiction to jurisdiction. The Property Transaction Network (www.theptn.com) is already offering an 'Electronic Closing Table' on which the real estate transaction can be completed online. This 'Table' provides a secure area in which all the transaction's participants, i.e. real estate brokers, insurers, title companies and escrow representatives, may safely exchange documents.

At present, the developed MLS does not allow for the performance of the following functions: multiple criteria analysis of alternatives (priority, utility degree and market value of the analysed real estate alternatives); negotiations and determination of the most rational real estate purchase variant based on real calculations.

The Real Estate's Market Value and Pollution and Health Effects Analysis Decision Support System (Zavadskas et al., 2007) and the Cooperative Integrated Web-Based Negotiation and Decision Support System for Real Estate (Kaklauskas et al., 2005) developed by the authors create conditions for e-listing, e-searching, e-evaluation, e-negotiations, e-execution and above functions. For example, Real Estate's Market Value, Pollution and Health Effects Analysis Decision Support System consists of Market Value Analysis, Air Pollution, Premises Microclimate Analysis, Health Effects, Voice Stress Analysis, Cooperative Decision Making and Multiple User subsystems.

Neural networks, expert and decision support systems and their integration

An expert system is a computer programme, or set of computer programmes, that contains a knowledge base and a set of rules that infer new facts from the knowledge and from the incoming data and are then used to help solve problems in certain areas. Moreover, the system performs many secondary functions, as an expert does, such as asking relevant questions, explaining its reasons and the like. The degree of problem solving is based on the quality of the data and the rules. Expert systems today generally serve to relieve a 'human' professional of some difficult, but clearly formulated, tasks.

A decision support system (DSS) is an information system that stores and processes information and data from various sources. By using different mathematical and logical models, it provides the decision maker with the necessary information for analysing, compiling and evaluating possible decision alternatives, and for making decisions and effecting the output and storage of the obtained results. Therefore, the DSS, which can be based on the data accumulated from different sources, should enable consumers to transform a huge amount of unprocessed data into information necessary for the analysis of a particular problem and for further decision making. DSS provides a framework through which decision makers can obtain the necessary assistance for making a decision through an easy-to-use menu or command system. Generally, a DSS will provide help in formulating alternatives, accessing data, developing models and interpreting their results, selecting options, or analysing the impacts of a selection.

A neural network is a method of computing that tries to copy the way a human brain works. A group of processing elements receives data at the same time and links are made between the elements, as repeated patterns are recognized (Oxford Dictionary of Computing, 1996). Many various-purpose neural networks, expert and decision support systems can be used for investment analysis, investment performance, portfolio analysis, management of investment, comfort zone, land use regulation, real estate development, risk, valuation, real estate transaction, negotiation, financing etc.

Integration of neural networks, multimedia, knowledge-based, decision support and other systems in real estate investment has a very promising future in scientific research. Recently, much effort has been made in order to apply the best elements of multimedia, neural networks and knowledge-based and other systems to DSSs.

Knowledge-based and decision support systems are related, but they treat decisions differently. For example, knowledge systems are based on previously obtained knowledge and rules of problem solving, whereas a DSS leaves quite a lot of space for a user's intuition, experience and outlook. Knowledge systems form a decision trajectory themselves, while DSSs perform a passive auxiliary role, though a situation might occur when DSSs suggest further actions to the decision maker. Calculation and analytical DSS models can be applied to process information and knowledge that is stored in the knowledge base. For example, some DSS models can be applied to prepare recommendations by referring to the knowledge in the knowledge base. DSSs can also facilitate the search and an analysis and distribution of the explicit knowledge.

Some think that computer (i.e. agent) intermediaries will replace human intermediaries. This is unlikely, as they have different qualities and abilities. It is far more likely that computer intermediaries will cooperate closely with humans, and that there will be a shift in the tasks (i.e. queries) that both types handle. Computer agents, in the short and medium term, will handle standard tasks and all those tasks that a computer programme (i.e. an agent) can do faster or better than a human can. Human intermediaries will handle very complicated problems, and will divide these tasks into subtasks that can, but not necessarily have to, be handled by intermediary agents. It may also be expected that many commercial parties, e.g. human information brokers, publishers etc., will want to offer middle layer services (Hermans, 2000).

References

American Lung Association (2005) *State of the Air*. American Lung Association.

Bardwick, J. (1995) *Danger in the Comfort Zone: How to Break the Entitlement Habit that's Killing American Business*. American Management Association: New York. ISBN 0–8144–7886–7

Cohen, A.J., Ross Alexander, H., Ostro, B. et al. (2005) The global burden of disease due to outdoor air pollution. *Journal of Toxicology and Environmental Health*, 68, 1–7.

Diversification. Poole: Moneyterms Home. Available from: http://moneyterms.co.uk/ diversification/ (accessed 24 December 2007).

Dhillon, U.S., Upinder, S., Schilling, J.D. and Sirmans, C.F. (1990) The mortgage maturity decision: the choice between 15-year and 30-year FRMs. Southern Economic Journal, 56, 1103–1116.

Efficient frontier. Poole: Moneyterms Home. Available from: http://moneyterms.co.uk/ efficient-frontier/ (accessed 24 December 2007).

Efficient portfolio. Poole: Moneyterms Home. Available from: http://moneyterms.co.uk/ efficient-portfolio/ (accessed 24 December 2007).

Expected return. Poole: Moneyterms Home. Available from: http://moneyterms.co.uk/ expected-return/ (accessed 24 December 2007).

Grass, D. Available from: http://davegrass.com/ (accessed 28 December 2007).

Ham, S. and Atkinson, R.D. (2003) *Modernizing Home Buying How IT Can Empower Individuals, Slash Costs, and Transform the Real Estate Industry*. Policy Report, Progressive Policy Institute. p. 21.

Hermans, B. (2000) Intelligent Software Agents on the Internet: an inventory of currently offered functionality in the information society and a prediction of (near-) future developments. http://www.hermans.org/agents/index.html (accessed 28 December 2007).

Kaklauskas, A., Zavadskas, E. and Andruskevicius, A. (2005) Cooperative Integrated Web-Based Negotiation and Decision Support System for Real Estate. *Lecture Notes in Computer Science, Cooperative Design, Visualization, and Engineering*, Volume 3675/2005.

Lam, K.C., Runeson, G., Tam, C.M. and Lo., S.M. (1998) Modelling loan acquisition decisions. *Engineering, Construction and Architectural Management*, 5(4), 359–375.

Leece, D. (1997) Mortgage design in the 1990s: theoretical and empirical issues. *Journal of Property Finance*, 8(3), 226–245.

Leece, D. (2000) Choice of mortgage instrument, liquidity constraints and the demand for housing debt in the UK. *Applied Economics*, 32, 1121–1132.

Markowitz, H.M. (1952) Portfolio selection. *Journal of Finance*, 7(1), 77–91.

Mortgage. Poole: Wikipedia. Available from: http://en.wikipedia.org/wiki/Mortgage (accessed 27 December 2007).

Oxford Dictionary of Computing (1996) Oxford University Press. p. 394.

Patrick, H. Available from: http://www.natdir.com/ (accessed 28 December 2007).

Portfolio theory. Poole: Capital Performance partners. Available from: http:// www.cperformance.com/Glossary.htm (accessed 25 December 2007).

Portfolio theory. Poole: Moneyterms Home. Available from: http://moneyterms.co.uk/ portfolio-theory/ (accessed 24 December 2006).

Quadreal (2007) How to build an investment comfort zone. http://www.quadreal.com/ resource_newsletter11.html (accessed 24 December 2007).

Rasmussen, D.W., Megbolugbe, I.F. and Morgan, B.A. (1997) The revenue mortgage as an asset management tool. *Housing Policy Debate*, 8(1), 173–194.

Real Estate Offer Generator (REOG). Poole: Super shareware.com. Available from: http://www.supershareware.com/download/reog-real-estate-offer-generator.html (accessed 26 December 2007).

Real Estate Offer Generator. Poole: Program URL.com. Available from: http://

www.programurl.com/software/investment-portfolio5.htm (accessed 27 December 2006).

Real Estate Notebook. Poole: Download 32. Available from: http://www.download 32.com/real-estate-notebook-i6972.html (accessed 27 December 2007).

Real Estate Tracker. Poole: Portfolio software. Available from: https://mydomuspro.com/portfolio-software.htm (accessed 26 December 2007).

Renaud, B. (1999) The financing of social housing in integrating financial markets: a view from developing countries. *Urban Studies*, 36(4), 755–773.

RLI, Real estate investing. Available from: http://en.wikipedia.org/wiki/Real_estate_investing

Schwartz, J. G. Available from: http://www.jgschwartz.com/CM/FSDP/PracticeCenter/Real-Estate/Real-Estate.asp?focus=topic&id=1 (accessed 28 December 2007).

Software for real estate investment. Poole: Bigger pockets forums. Available from: http://forums.biggerpockets.com/post-49610.html (accessed 25 December 2007).

Van Dyk, N. (1995) Financing social housing in Canada. *Housing Policy Debate*, 6(4), 815–848.

Volatility. Poole: Moneyterms Home. Available from: http://moneyterms.co.uk/volatility/ (accessed 24 December 2007).

Wallmeier, M. and Zainhofer, F. (2006) How to invest over the life cycle: Insights from theory. *Journal für Betriebswirtschaft*, 56, 4.

Zavadskas, E., Kaklauskas, A., Maciunas, E. et al. (2007) Real Estate's Market Value and a Pollution and Health Effects Analysis Decision Support System. *Lecture Notes in Computer Science, Cooperative Design, Visualization, and Engineering*, Volume 4674/2007.

Index